MODERN ENGINEERED BAMBOO STRUCTURES

PROCEEDINGS OF THE SUSTAINABLE BAMBOO BUILDING MATERIALS SYMPOSIUM OF BARC 2018 AND THE 3RD INTERNATIONAL CONFERENCE ON MODERN BAMBOO STRUCTURES (ICBS 2018), BEIJING, CHINA, 25-27 JUNE 2018

Modern Engineered Bamboo Structures

Editors

Y. Xiao

Zhejiang University – University of Illinois at Urbana Champaign Institute (ZJUI), Zhejiang University International Campus in Haining, China

Z. Li

College of Civil Engineering, Nanjing Tech University, China

K.W. Liu

International Bamboo and Rattan Organisation, Beijing, China

CRC Press is an imprint of the
Taylor & Francis Group, an **informa** business
A BALKEMA BOOK

Published by:
CRC Press/Balkema
Schipholweg 107C, 2316 XC Leiden, The Netherlands

First issued in paperback 2023

ISBN 13: 978-1-03-257059-4 (pbk)
ISBN 13: 978-1-138-35185-1 (hbk)
ISBN 13: 978-0-429-43499-0 (ebk)

DOI: https://doi.org/10.1201/9780429434990

Visit the Taylor & Francis Web site at
http://www.taylorandfrancis.com

and the CRC Press Web site at
http://www.crcpress.com

Typeset by Integra Software Services Pvt. Ltd., Pondicherry, India

Library of Congress Cataloging-in-Publication Data

Publisher's Note
The publisher has gone to great lengths to ensure the quality of this reprint but points out that some imperfections in the original copies may be apparent.

Applied for

Modern Engineered Bamboo Structures – Xiao, Li & Liu (eds)
© 2020 Taylor & Francis Group, London, ISBN 978-1-138-35185-1

Table of contents

Modern Engineered Bamboo Structures – Xiao, Li & Liu (eds)
© 2020 Taylor & Francis Group, London, ISBN 978-1-138-35185-1

Preface

The 3rd International Conference on Modern Bamboo Structures (ICBS-2018) was held in Beijing, China, June 25-17, 2018, following the previous successful events in Changsha, China, 2007, and in Bogota, Columbia, 2009, and several subsequent workshops. The 3-ICBS-2018 was organized as an important thematic event of the 2018 Global Bamboo and Rattan Congress (BARC 2018 http://www.barc2018.org/en), aiming at providing an open forum for experts around the world to exchange information and to discuss topics related to design, analysis, testing, manufacture, construction and education of modern bamboo structures.

The conference featured nine presentation sessions, panel discussions, exhibitions, etc., and was attended by more than 200 participants, representing a wide spectrum of professions including academia, architecture, engineering, manufacture, government, as well as NGOs. After the conference, the authors are solicited to submit their papers and more than twenty papers are selected to be published in the ICBS-2018 Proceedings, after a peer review process. The topics of the papers cover but not limited to the followings,

- Properties of structural bamboo materials, components
- Composites of bamboo, timber and other materials
- Adhesives, joints, etc.
- Engineered bamboo products
- Hybrid systems with composite of bamboo and other materials
- Architectural and structural design concepts for modern bamboo structures
- Lightweight frame systems with bamboo or timber
- Practices of modern bamboo buildings and bridges, etc.

Several state-of-the-art reports are also included in the proceedings, in the hope to provide a tracking records about the development of modern bamboo structures in the past ten years.

The editors would like to acknowledge Inbar and many co-sponsoring organizations, collaborators for their supports to the success of the ICBS-2018 conference. We thank all the participants, presenters, and authors for their contributions to the development of modern bamboo structures. It is our hope that the proceedings can serve as a reference to shed lights to the future trend of engineered design and industrialization of bamboo materials and structures.

Y. Xiao, Z. Li and K.W. Liu
Editors

Modern Engineered Bamboo Structures – Xiao, Li & Liu (eds)
© 2020 Taylor & Francis Group, London, ISBN 978-1-138-35185-1

Editor biographies

Dr. Y. Xiao is Distinguished Professor, Program Director for Energy, Environment and Infrastructure Sciences, in the Zhejiang University – University of Illinois at Urbana Champaign Institute (ZJUI), Zhejiang University International Campus in Haining, China. His academic career spans in Japan, US and China. He serves as the associate editor for the ASCE Journal of Structural Engineering, Journal of Bridge Engineering, and editorial board member of the Journal of Constructional Steel Research. He is elected fellows of the American Society of Civil Engineers (ASCE) and the American Concrete Institute (ACI). Passionate about bamboo, Dr. Xiao focuses and publishes pioneered research on engineered bamboo and holds many patents and the trademark technology GluBam®. In 2008, he was awarded the Best of What's New in 2008 by the Popular Science Magazine, and was named as the Popular Science Innovator of 2008. Prof. Xiao's team has designed and built many demonstration bamboo houses and bridges in China and Africa.

Dr. Z. Li graduated in civil engineering in 2009 from the Southwest Jiaotong University, then was enrolled in master/doctoral courses at the Hunan University. As an exchange student, he pursued doctoral courses at the Sapienza University of Rome, through Erasmus project (Eurasian University Network for International Cooperation in Earthquakes), and obtained a Ph. D. degree there in 2015. From Feb. 2015 to June 2016, he worked as a postdoctoral researcher at Sapienza University of Rome. He is now on the faculty as an assistant professor in college of civil engineering, Nanjing Tech University. Since 2009, he has been working on modern bamboo structures and has published more than 12 international journal papers related to this topic.

Mrs. K.W. Liu is the Coordinator of Global Bamboo Construction Programme at the International Bamboo and Rattan Organisation (INBAR – www.inbar.int), based at its Headquarters in Beijing, China. Mrs. Liu has managed more than 10 international bamboo construction projects across Asia, Africa and Latin American in the last 10 years. She is also one of the founders of INBAR Construction Task Force (INBAR TFC) and currently manages a group of 28 bamboo construction experts from 18 countries. As the Convenor of ISO/TC165 (Timber Structures) Working Group (WG) 12 – Structural Uses of Bamboo, Mrs. Liu is leading the development and revision of several international and Chinese standards for bamboo construction. Mrs. Liu is one of experts of "Timber and Bamboo Structures Committee" and a member of "Sustainable Civil Engineering Committee" organized by Chinese Society for Urban Studies in China. As the first author, she published the book on Contemporary Bamboo Architecture in China. She also translated into Chinese the book Sustainable Building in Practice: What the Users Think written by George Baird, published by China Architecture & Building Press in 2019 and 2013 respectively.

Modern Engineered Bamboo Structures – Xiao, Li & Liu (eds)
© 2020 Taylor & Francis Group, London, ISBN 978-1-138-35185-1

INBAR and BARC 2018

INBAR

Established in 1997, the International Bamboo and Rattan Organization (INBAR) is a multilateral development organization that promotes the use of bamboo and rattan for sustainable development. It currently has 45 Member States. In addition to its Secretariat Headquarters in China, INBAR has Regional Offices in Cameroon, Ethiopia, Ecuador, India and Ghana.

The mission of INBAR is to improve the well-being of producers and users of bamboo and rattan within the context of a sustainable bamboo and rattan resource base, by consolidating, coordinating and supporting strategic and adaptive research and development.

INBAR's work is based around the following strategic goals:

1. Promoting bamboo and rattan in socio-economic and environmental development policies at national, regional and international levels;
2. Coordinating inputs on bamboo and rattan from a growing global network of Member States and partners, and representing the needs of Member States on the global stage;
3. Sharing knowledge and communicating lessons learned, providing training and raising awareness of the relevance of bamboo and rattan as plants and commodities;
4. Fostering adaptive research and on-the-ground innovation by promoting pilot case studies and supporting upscaling of best practices across INBAR Member States.

As an Observer to the United Nations General Assembly, INBAR is an important representative for its Member States, over 40 of which hail from the Global South. INBAR has played an especially strong role in promoting South-South cooperation over the last 20 years.

Since its founding in 1997, INBAR has been making a real difference to the lives of millions of people and environments around the world, with achievements in areas such as: raising standards; promoting safe, resilient bamboo construction; restoring degraded land; training up to 25,000 bamboo and rattan practitioners; and informing green policy and sustainable development objectives.

BARC 2018

The Global Bamboo and Rattan Congress, BARC 2018, opened on 25 June 2018 in Beijing, China. The three-day event, co-hosted by the International Bamboo and Rattan Organization and China's National Forestry and Grassland Administration, welcomed around 1200 participants from 70 countries, including ministers, policymakers and representatives from research institutes, development organizations, UN bodies and the private sector.

The event featured a ministerial summit, three high-level dialogues - which covered South-South cooperation, climate change, innovation and industry development - and around 80 parallel sessions. The overarching theme of the Congress was 'Enhancing South-South Cooperation for Green Development through Bamboo and Rattan's Contribution to the Sustainable Development Goals'. Discussions centered around how to realize bamboo and rattan's huge potential in a number of areas: sustainable commodity production, disaster-resilient construction, poverty alleviation, climate change mitigation and adaptation, land restoration and biodiversity protection.

'Critical, but underused' - this was the key message of BARC 2018. Fast-growing and local to some of the poorest communities in the tropics and subtropics, bamboo and rattan hold huge potential for climate change mitigation, innovative construction and job creation, but are rarely used for more than simple construction and household use.

From its inception, BARC 2018 was planned as a very results-focused event: a platform for sharing knowledge, promoting new partnerships and programs and inspiring participants to integrate bamboo and rattan into their policies, business plans and discussions. The main themes of BARC 2018 reflect the focus areas of INBAR's 2015 to 2030 Strategy: 'From Research to Development'. In particular, the Congress launched the Beijing Declaration. The Declaration recognizes bamboo and rattan's various benefits and commits 'ministers, senior officials and participants' to call upon national governments and other bodies to implement a number of recommendations. The Declaration marks a significant step for the development of bamboo and rattan.

可持续竹建材研讨会暨第三届现代竹结构国际会议

Sustainable Bamboo Building Materials Symposium & 3rd International Conference on Modern Bamboo Structure

26-27th June, 2018年6月26日-27日

Beijing, China 中国北京

State-of-art Reports

Modern Engineered Bamboo Structures – Xiao, Li & Liu (eds)
© 2020 Taylor & Francis Group, London, ISBN 978-1-138-35185-1

Recent progress in engineered bamboo development

Y. Xiao
Zhejiang University-University of Illinois Institute (ZJUI), Zhejiang University, Haining, Zhejiang, China

B. Shan
Hunan University, Changsha, Hunan, China

Z. Li
Nanjing Tech University, Nanjing, Jiangsu, China

ABSTRACT: This paper provides a state of the art summary on the innovations and development in engineered bamboo materials and manufacture, mostly in recent ten years. The definition of engineered bamboo structures is given first as all the structures with components using bamboo as the main materials. However, the main topics discussed in this paper is on engineered bamboo structures with modifying the original round bamboo culm and mainly in the form of bamboo fiber/bundle or strips reinforced composites. Three types of engineered bamboo, plybamboo, laminated bamboo and bamboo scrimbers can be categorized and discussed in the paper. The most recent development on cross laminated bamboo and timber or CLBT is also introduced.

A brief review shows that the scientific publications are increasing dramatically since the first conference on modern bamboo in 2007.

1 INTRODUCTION

The use of bamboo in construction of buildings and bridges is popular in many of the bamboo-growing regions in the world and has been practiced for centuries. However, bamboo construction has been limited mostly to the form of round culms until later half of last century. Buildings or bridges using original round culm bamboo are favored by many architects due to the natural appealing, they are not very efficient in terms of manufacturing output. For example, if an engineer specifies 100 bamboo culm columns to have an outside diameter of 100 mm and a wall thickness of 10mm for a building design, collecting bamboo culms with such dimensions, even within a certain margin of error, may be a difficult task, since a harvest of bamboo comes with many varieties of dimensions. Furthermore, the irregularities of a bamboo culm are also found within its sections and along its length. The sections may not be the same or may not even be round and its diameter typically becomes smaller towards its top. In any case, round bamboo structures can still be designed and built according to modern engineering theories methods. Such structures should also be categorized into modern structures.

In the later half of last century, rapid change emerged for more efficient use of bamboo materials. Through industrialized manufacturing processes and engineering design, laminated bamboo composite materials are developed. These engineered materials are made into regular shapes that can be designed and sized accurately for construction purposes, similar to other modern industrialized materials, such as modern timber. The products are also similar to those of wood, such as glued laminated lumber, glulam cross laminated timber or CLT, which are well-established industrial products used in construction, mostly in industrialized countries. Largely, there are mainly three types of engineered bamboo products manufactured using adhesive lamination, and they are summarized in this paper along with most recent products.

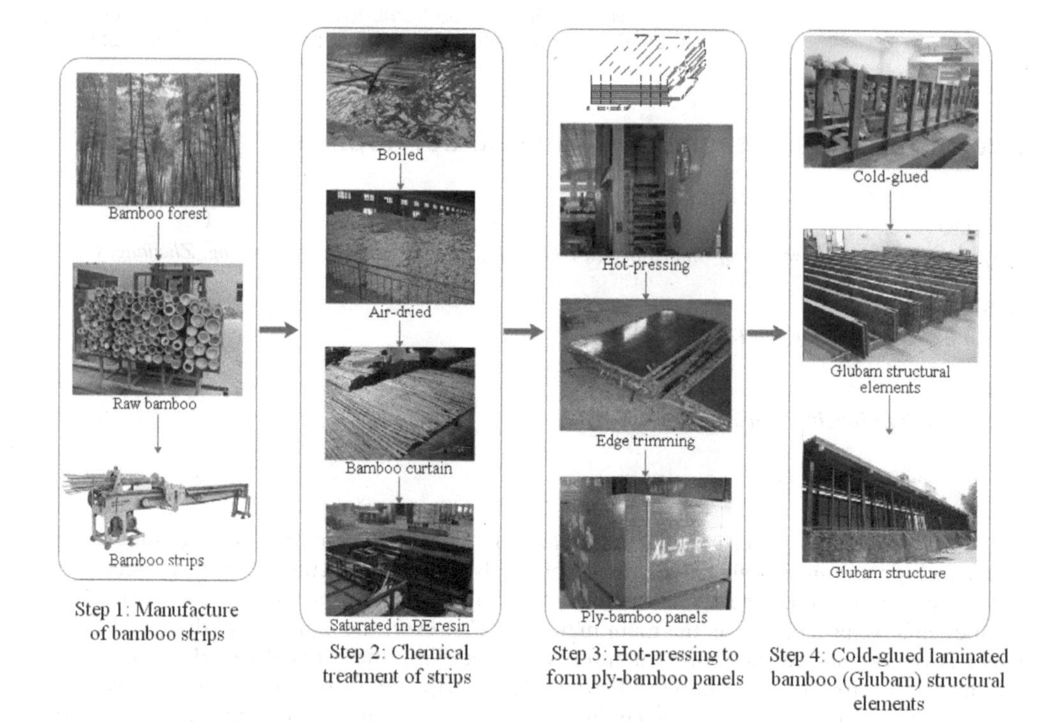

Figure 1. Production of Glubam structures based on plybamboo panels.

Similar to engineered wood, such as Glulam or Cross Laminated Timber structures, engineered bamboo products can also be categorized into hot and cold formation. Figure 1 shows the typical process of producing glue laminated bamboo (GluBam) structures.

The development of engineered bamboo products in recent years used for Glubam is discussed as following categories.

2 PLY-BAMBOO

Plywood is a great advancement in modern timber industry with lamination of thin wood-veneers. Similar to plywood, plybamboo can be manufactured using a procedure with a few steps similar to that used for manufacturing plywood. There are two typical types of plybamboo sheets, which can be categorized as the thin layer lamination and thick layer lamination, as shown in Figure 2.

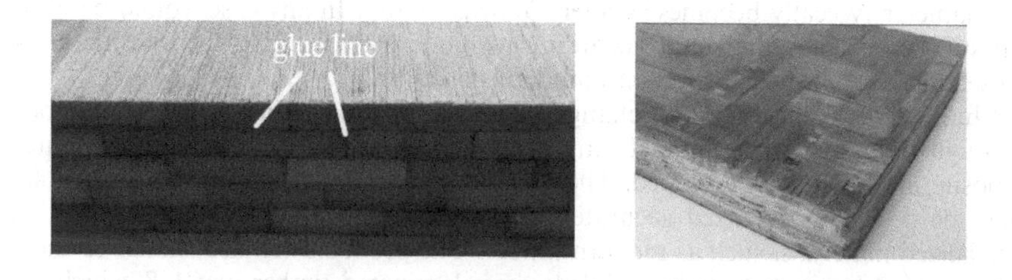

(a) plybamboo made from thick bamboo strips (b) plybamboo made from thin bamboo strips

Figure 2. Ply-bamboo made with different bamboo strips.

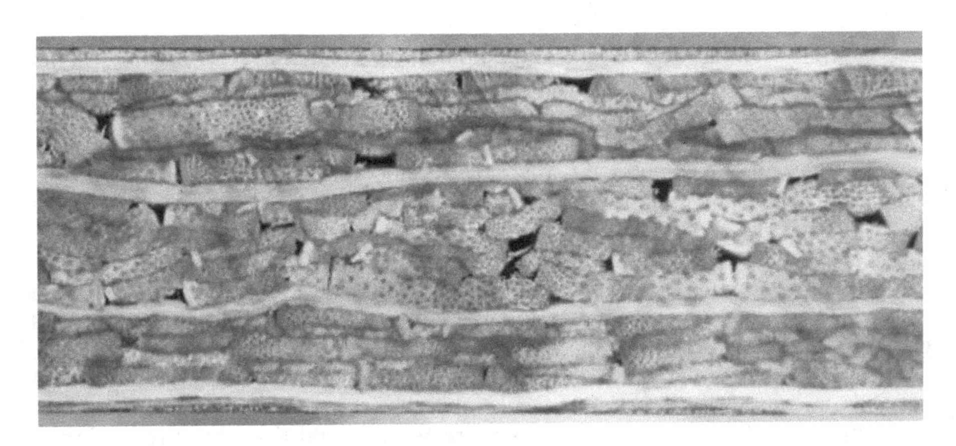

Figure 3. Section of a recently manufactured plybamboo.

The thick layer laminated bamboo sheets are made by pressure gluing a few layers (typically three to five layers) of relatively thicker (about 5~8 mm) bamboo strips with a width of about 20~25 mm. The top of the line products are made for the use as flooring plates, as alternative to wood flooring. The thin layer laminated bamboo sheets typically have a thickness of about 10 to 30 mm, and are made by laminating approximately 2 mm thick bamboo strip mats. They are mass produced and mainly used as concrete formwork or flooring plates for buses, trucks and container boxes in China. The strips are netted into mats which can be prepared by local farmers before bringing to a factory. At the factory, the mats are cleaned, dried in a kiln, and saturated in resin. The resin saturated bamboo strip mats need to be dried again and finally stacked and pressed under a temperature of 150°C, using a procedure similar to manufacturing plywood. The cost for making plybamboo using thinner layer bamboo strips is relatively low due to the fact that it makes full use of bamboo culms of varieties of geometry.

In the past, the quality of the thin layer plybamboo boards was varied from manufacturers to manufactures, particularly, the thickness could not be well controlled to the ideal accuracy, and often could vary for 10%. A recent innovation is to add a few wood layers, as shown by the lighter layers in the section view of a thin-layer plybamboo shown in Figure 3. Due to the fact that wood is more compressible than bamboo, the wooden layers serve as cushions during the hot-pressing process and makes it easier to control the uniform thickness. Meanwhile, the newly produced plybamboo eliminated the need for removing the bamboo bark and the inner layers, greatly increased the usage of bamboo mass.

3 LAMINATED BAMBOO – LBL AND GLUBAM

Laminated bamboo can be considered as a general terminology for bamboo products processed using adhesive lamination. In this sense, of course, plybamboo is also a type of laminated bamboo. However, in this section, we mainly define laminated bamboo as the components that can be directly used as structural components.

One of the laminated bamboo products is abbreviated as LBL (laminated bamboo lumber), which is manufactured using a hot pressing process. Single layers are typically made first, and then the layers are pressed together to form the desired size of structural elements, under an elevated temperature of 140±5°C. During the pressing process, a perpendicular to plane pressure of about 5 MPa and a transverse pressure of about 2 MPa are applied. Due to the gradient loss in thermal transfer, the total thickness of the components can be made with a hot pressing process is limited.

Another laminated bamboo product is named glubam, or glued laminated bamboo, a term mimicking the well-established timber product glulam. Glubam was originally trademarked by Xiao in the USA, and now is almost a commonly used technical terminology similar to glulam. Generally, glubam is defined as the laminated bamboo with a two-step pressing lamination process. The first step is the production of plybamboo with the hot pressing process. The second step is a pressing process normally under room temperature, also known as cold pressing, in which the plybamboo are sized into specific dimension according to design need and then glued under pressure into the required structural elements, as the example shown in Figure 4.

Analogically, one can compare the production of glubam with steel structures: the first step of making glubam is the plybamboo production, like the rolling of steel plates or sheets, while the second step is the forming of specific structural elements based on the design requirements, using procedures like cutting and welding (adhesive lamination for glubam) for steel structure. Compared with the hot pressing process-based LBL, the advantage of glubam is its relative cost effectiveness, because the transportation of the plates and sheets produced in the first step process is more efficient and the second step of the cold-pressing process can be easily established, using similar equipment like those for producing timber based glulam.

Structural components of glubam, such as beams and columns can be connected using similar connection details as used for glulam structures, typically, involving the use of steel connection plates and bolts. Figure 5 shows the connection plate details and an example of glubam frame under installation.

Figure 4. Room condition formation of glubam girders.

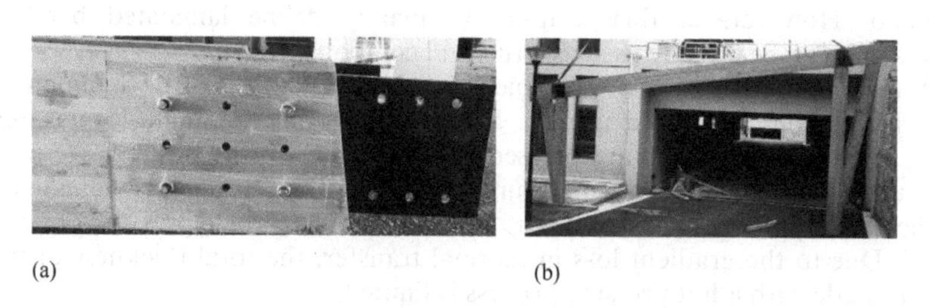

(a) (b)

Figure 5. Glubam frame details.

4 BAMBOO SCRIMBERS

Bamboo scrimber is a relatively new type of engineered bamboo with adhesive impregnated bamboo bundles (fibers) densified to form lumber-like components. The process is somewhat similar to wood scrimbers, mainly includes cold molding, hot curing and hot-pressing. Normally the relative density of bamboo scrimber exceeds that of water, or 1.0, with final density of twice of original bamboo or more, reaching about 1200 kg/m^3 or higher. The density and the hardness of bamboo scrimber make it difficult to be used as an alternative of timber in construction, unless using procedure of metal manufacturing of cutting and drilling, etc. Most of the bamboo scrimber products are mainly used as outdoor decorations for wall panels, and flooring decks. However, a recent trade-marked product, Bamboo Steel, is also being pushed into construction of buildings, with components pre-manufactured and pre-formed for site-installation. Figure 6 shows a recent example building constructed using so-called bamboo steel.

Figure 6. Building constructed with a trade-marked bamboo scrimber-bamboo steel (courtesy to Mr. G. Wang).

5 CLBT-CROSS-LAMINATED BAMBOO AND TIMBER

Inspired by the recent development of CLT, or cross-laminated timber, the first author developed cross-laminated bamboo and timber, or CLBT (can also be named as CLTB). The CLBT is made by glue laminating under pressure the layers of glubam board and standard timber lumber elements. The orientation of the lumbers can be predetermined to satisfy design requirement. Figure 7 shows the examples of CLBT. The author's team has

Figure 7. An example of CLBT with thin-layer plybamboo as surfaces and SPF as core.

successfully made CLBT with the combinations of SPF and popular lumbers with different glubams.

6 INTERNATIONAL TREND ON BAMBOO RELATED RESEARCH

Since the first international conference on modern bamboo structures in 2007 (Xiao et al. 2008), there is a steady increase in scientific journal publications of bamboo related research. All the papers are listed in Appendix of this paper. Figure 8 depicts the annual numbers of SCI indexed papers in last ten years. It should be clarified that most of the papers are mainly searched from the academic journals related to structural materials, structures and construction, thus may be limited in scope. The number of the papers published in the year of 2018 is slightly lower may be due to the fact that some publications have not been indexed.

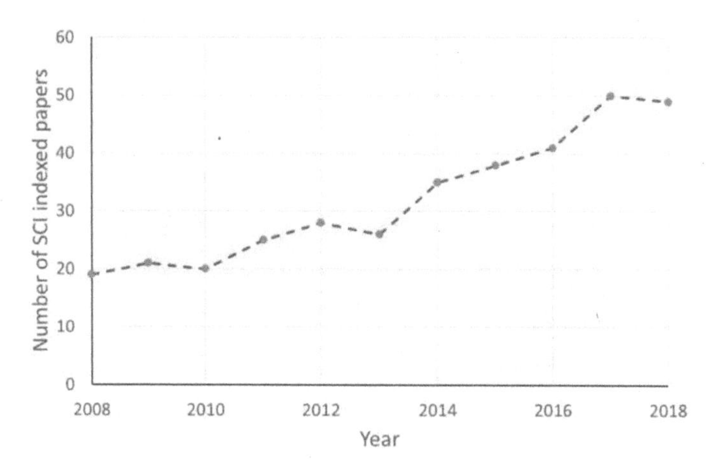

Figure 8. Annual journal papers related to bamboo structures in last ten years.

7 CONCLUSIONS

Since the first international workshop of modern bamboo structures in 2007, in only ten years, we have seen many innovations in bamboo products for applications in structures. Meanwhile, many organizations, entrepreneurs have joined the efforts to develop new and better bamboo products and structures.

Based on a limited review of international journal publications, the scientific research on bamboo based structures is increasing steadily in last ten years, since the first International Conference of Modern Bamboo Structures.

ACKNOWLEDGEMENTS

The authors greatly appreciate the supports of various sponsoring agencies and organizations for their enthusiasm in sustainable development and the cultivation of bamboo products.

REFERENCES

De Capua A., (2015), Towards global architecture. The project between technique and technology, n.14 (2)/2015, in Budownictwo i Architektura.

De Capua A., Ciulla V. (2017), Osservatorio P.A.R.C.O. Caratterizzazioni per la qualità ambientale indoor, TECHNE, vol. 1, p. 209–217.

Li, Z., Xiao, Y., Wang, R., & Monti, G. (2014). Studies of nail connectors used in wood frame shear walls with ply-bamboo sheathing panels. *Journal of Materials in Civil Engineering*, 27(7), 04014216.

Piatkowska, K.K. (2013). Expo Pavilions As Expression Of National Aspirations. Architecture As Political Symbol. Archhist' 13 Architecture Politics Art Conference, Politics in the history of architecture as cause & consequence, Vol. 48, 2013, pp. 20–29.

Tornatora, M., (2017), Learning from pavilion: 100+ 100. Gangemi Editore.

Wang, J.S., Demartino, C., Xiao, Y., & Li, Y.Y. (2018). Thermal insulation performance of bamboo-and wood-based shear walls in light-frame buildings. *Energy and Buildings*, 168, 167–179.

Xiao, Y., & Ma, J. (2012). Fire simulation test and analysis of laminated bamboo frame building. *Construction and building materials*, 34, 257–266.

Xiao, Y., Li, Z., & Wang, R. (2014). Lateral loading behaviors of lightweight wood-frame shear walls with ply-bamboo sheathing panels. *Journal of Structural Engineering*, 141(3), B4014004.

Xiao, Y., Paudel, S.K., and Inoue, M. (2008). "Modern bamboo structures." London, UK, Boca Raton: CRC Press.

Xiao, Y., Zhou, Q., & Shan, B. (2009). Design and construction of modern bamboo bridges. *Journal of Bridge Engineering*, 15(5), 533–541.

APPENDIX-SCI indexed papers published in 2008–2018

Chang, H.C., Lee, K.H., Kim, J.S., & Kim, Y.S. (2008). Micromorphological characteristics of bamboo (phyllostachys pubescens) fibers degraded by a brown rot fungus (gloeophyllum trabeum). Journal of Wood Science, 54(3), 261.

Hiziroglu, S., Jarusombuti, S., Bauchongkol, P., & Fueangvivat, V. (2008). Overlaying properties of fiberboard manufactured from bamboo and rice straw. Industrial Crops and Products, 28(1), 107–111.

Hui, W., Rui, C., Sheng, K.C., Adl, M., & Qian, X.Q. (2008). Impact response of bamboo-plastic composites with the properties of bamboo and polyvinylchloride (pvc). Journal of Bionic Engineering, 5(08), 28–33.

Li, Z. (2008). The thermal residual stress redistribution in a bamboo interconnect. Journal of Thermal Stresses, 31(7), 638–648.

Lima, H.C., Willrich, F.L., Barbosa, N.P., Rosa, M.A., & Cunha, B.S. (2008). Durability analysis of bamboo as concrete reinforcement. Materials & Structures, 41(5), 981–989.

Lin, C.M. &Chang, C.W. (2008). Production of thermal insulation composites containing bamboo charcoal. Textile Research Journal, 78(7), 555–560.

Ma, J.F., Chen, W.Y., Zhao, L., & Zhao, D.H. (2008). Elastic buckling of bionic cylindrical shells based on bamboo. Journal of Bionic Engineering, 5(3), 231–238.

MahuyaDas, & DebabrataChakrabarty. (2008). Study of impact properties and morphology of unidirectional bamboo strips–polyester composites: effect of mercerization. Composite Interfaces, 15(7–9), 11.

Mui, E.L.K., Cheung, W.H., Lee, V.K.C., & Mckay, G. (2008). Kinetic study on bamboo pyrolysis. Industrial & Engineering Chemistry Research, 47(15), 5710–5722.

Ogihara, S., Okada, A., & Kobayashi, S. (2008). Mechanical properties in a bamboo fiber/pbs biodegradable composite. Journal of Solid Mechanics & Materials Engineering, 2(3), 291–299.

Pande, S.K., & Pandey, S. (2008). Bamboo for the 21st century. International Forestry Review, 10(2), 134–146.

Shan, W., & Li, Y. (2008). Application prospect of bamboo utilization in building structures. Forest Engineering.

Shibata, S., Yong, C., & Fukumoto, I. (2008). Flexural modulus of the unidirectional and random composites made from biodegradable resin and bamboo and kenaf fibres. Composites Part A Applied Science & Manufacturing, 39(4), 640–646.

Sumardi, I., Kojima, Y., & Suzuki, S. (2008). Effects of strand length and layer structure on some properties of strandboard made from bamboo. Journal of Wood Science, 54(2), 128–133.

Sun, Z., Cheng, Q., & Jiang, Z. (2008). Processing and properties of engineering bamboo products. Acta Materiae Compositae Sinica.

Sungkaew, S., Teerawatananon, A., Parnell, J.A.N., Stapleton, C.M.A., & Hodkinson, T.R. (2008). Phuphanochloa, a new bamboo genus (poaceae: bambusoideae) from thailand. Kew Bulletin, 63(4), 669–673.

Xiao, Y., Shan, B., Chen, G., Zhou, Q., & She, L.Y. (2008). Development of a new type of glulam—Glubam. Modern Bamboo Structures.

Yang, G.S., Zhang, Y.P., Shao, H.L., Hu, X.C., & Rosenau, T. (2008). A comparative study of bamboo lyocell fiber and other regenerated cellulose fibers. Holzforschung, 63(1), 18–22.

Zhang, X., & Zhang, F. (2008). A study on the "hollow design" of chinese bamboo product. IEEE.

Chen, X., Zhang, X., Zhang, Y., Booth, T., & He, X. (2009). Changes of carbon stocks in bamboo stands in china during 100 years. Forest Ecology and Management, 258(7), 0–1496.

Chugh, S., Tuli, S., Chatterjee, K., Singh, S., & Sudhakar, P. (2009). Bamboo-glue interface thermography for non-destructive testing. Journal of Bamboo & Rattan, 8(3/4), 143–148.
Corradi, S., Isidori, T., Corradi, M., Soleri, F., & Olivari, L. (2009). Composite boat hulls with bamboo natural fibres. International Journal of Materials and Product Technology, 36(1–4), 73–89.
Dan-Ke, D.U., Meng-You, J., Jiong, H., & Xin-Chang, Z. (2009). Research of elastic modulus of mould pressing bamboo processing fragment plate. Packaging Engineering, 30(8), 48–50.
Jarusombuti, S., Hiziroglu, S., Bauchongkol, P., & Fueangvivat, V. (2009). Properties of sandwich-type panels made from bamboo and rice straw. Forest Products Journal, 59(10), 52–57.
Kim, H., & Fujii, T. (2009). Study on tensile properties and Interfacial Shear Strength of bamboo fibre bundles.International Journal of Materials and Product Technology, 36(1–4), 115–124.
Kushwaha, P., & Kumar, R. (2009). Enhanced mechanical strength of bfrp composite using modified bamboos. Journal of Reinforced Plastics & Composites, 28(23), 2851–2859.
Manalo, R.D., & Acda, M.N. (2009). Effects of hot oil treatment on physical and mechanical properties of three species of Philippine bamboo. Journal of Tropical Forest Science, 19–24.
Ohmae, Y., Saito, Y., Inoue, M., & Nakano, T. (2009). Mechanism of water adsorption capacity change of bamboo by heating. European Journal of Wood and Wood Products, 67(1), 13–18.
Ohmae, Y., Saito, Y., Inoue, M., & Nakano, T. (2009). Water adsorption process of bamboo heated at low temperature. Journal of wood science, 55(1), 13–17.
Paes, J.B., Oliveira, A.K.F.D., Oliveira, E.D., & Lima, C.R.D. (2009). Caracterização físico-mecânica do laminado colado de bambu (dendrocalamus giganteus). Ciência florestal, 19(1).
Qin, L.I., Jian, Z., Kui-Hong, W., Bo, W., & Shao-Fei, Y. (2009). Technology of hot pressing for new composite bamboo plywood. Journal of Zhejiang Forestry Science & Technology, 29(5), 1–4.
Qin, L., & Yu, W. (2009). Status and prospects of reconstituted bamboo lumber. World Forestry Research, 22(6), 55–59.
Sulastiningsih, I.M., & Nurwati. (2009). Physical and mechanical properties of laminated bamboo board. Journal of Tropical Forest Science, 246–251.
Xiao, Y., She, L.Y., Shan, B., Zhou, Q., Chen, G., & Yang, R.Z. (2009). Application of modern bamboo struture to reconstruction after Wenchuan earthguake. Journal of Natural Disasters, 18(3), 14–18.
Xiao, Y., Zhou, Q., & Shan, B. (2009). Design and construction of modern bamboo bridges. Journal of Bridge Engineering, 15(5), 533–541.
Yan, X., Liyong, S., Bo, S., Quan, Z., Ruizhen, Y., & Guo, C. (2009). Research and design of prefabricated bamboo house. Industrial Construction, 39(1), 56–59.
Yang, G., Zhang, Y., Shao, H., & Hu, X. (2009). A comparative study of bamboo Lyocell fiber and other regenerated cellulose fibers 2nd ICC 2007, Tokyo, Japan, October 25–29, 2007. Holzforschung, 63(1), 18–22.
Yang, K.S., & Lee, S. (2009). The Analysis of the Mechanical Characteristic of Bamboo Net. Journal of the Korean Geotechnical Society, 25(5), 29–37.
Yeh, M.C., Hong, W.C., & Lin, Y.L. (2009). Flexural properties of structural laminated bamboo/solid wood composite box hollow beams. Taiwan Journal of Forest Science, 24(1), 41–49.
Zhang, Y.M., Yu, Y.L., & Yu, W.J. (2009). Effect of heat treatment on mechanical properties of phyllostachys pubescens bamboo.Transactions of Materials and Heat Treatment, 30(3), 33–35, 41.

Aschheim, M., Gil-Martín, L.M., & Hernández-Montes, E. (2010). Engineered bamboo I-joists. Journal of structural engineering, 136(12), 1619–1624.
Bonilla, S.H., Guarnetti, R.L., Almeida, C.M.V.B., & Giannetti, B.F. (2010). Sustainability assessment of a giant bamboo plantation in brazil: exploring the influence of labour, time and space. Journal of Cleaner Production, 18(1), 83–91.
Braga Filho, A.C., Lima Júnior, H.C., Barbosa, N.P., & Willrich, F.L. (2010). Structural behavior of concrete beams reinforced with pinned bamboo-splints. Revista Brasileira de Engenharia Agrícola e Ambiental, 14(10), 1115–1122.
Correal, J.F.D., & Juliana Arbeláez. (2010). Influence of age and height position on colombian "guadua angustifolia" bamboo mechanical properties. Maderas: Cienciay Tecnologia, 12(2), 105–113.
Correal, J.F., & Ramirez, F. (2010). Adhesive bond performance in glue line shear and bending for glued laminated guadua bamboo. Journal of Tropical Forest Science, 433–439.
Garrecht, H., Schneider, J., Ott, A., Franz, J., & Heinsdorff, M. (2010). Increase of the bearing load of the joints for the bamboo tubes of the German-Chinese House of EXPO 2010 by optimization of the bamboo-concrete bond. BAUINGENIEUR, 85, 353–360.
Huang, Y., Liu, H., He, P., Yuan, L., Xiong, H., Xu, Y., & Yu, Y. (2010). Nonisothermal crystallization kinetics of modified bamboo fiber/PCL composites. Journal of applied polymer science, 116(4), 2119–2125.

Hung, K.C., & Wu, J.H. (2010). Mechanical and interfacial properties of plastic composite panels made from esterified bamboo particles. Journal of Wood Science, 56(3), 216–221.
Kubojima, Y., Inokuchi, Y., Suzuki, Y., & Tonosaki, M. (2010). Shear modulus of several kinds of japanese bamboo obtained by flexural vibration test. Journal of Wood Science, 56(1), 64–70.
Lima Júnior, Humberto C., Willrich, Fábio L., Fabro, G., Rosa, M.A., Tanabe, L., & Sabino, René B.G. (2010). Análise mecânica de pilares mistos bambu-concreto. Revista Brasileira de Engenharia Agrícola e Ambiental, 14(5), 545–553.
Majumdar, A., Mukhopadhyay, S., & Yadav, R. (2010). Thermal properties of knitted fabrics made from cotton and regenerated bamboo cellulosic fibres. International Journal of Thermal Sciences, 49(10), 2042–2048.
Raghavendra Rao, H., Varada Rajulu, A., Ramachandra Reddy, G., & Hemachandra Reddy, K. (2010). Flexural and compressive properties of bamboo and glass fiber-reinforced epoxy hybrid composites. Journal of Reinforced Plastics and Composites, 29(10), 1446–1450.
Rana, M.P., Mukul, S.A., Sohel, M.S.I., Chowdhury, M.S.H., Akhter, S., & Chowdhury, M.Q., et al. (2010). Economics and employment generation of bamboo-based enterprises: a case study from eastern bangladesh. Small-scale Forestry, 9(1), 41–51.
Shao, Z.P., Fang, C.H., Huang, S.X., & Tian, G.L. (2010). Tensile properties of moso bamboo (phyllostachys pubescens) and its components with respect to its fiber-reinforced composite structure. Wood Science and Technology, 44(4), 655–666.
Shao, Z.P., Zhou, L., Liu, Y.M., Wu, Z.M., & Arnaud, C. (2010). Differences in structure and strength between internode and node sections of moso bamboo. Journal of Tropical Forest Science, 133–138.
Tsubaki, T., & Nakano, T. (2010). Creep behavior of bamboo under various desorption conditions. Holzforschung, 64(4), 489–493.
Wang, X.Q., Li, X.Z., & Ren, H.Q. (2010). Variation of microfibril angle and density in moso bamboo (phyllostachys pubescens). Journal of Tropical Forest Science, 22(1), 88–96.
Wong, K.J., Zahi, S., Low, K.O., & Lim, C.C. (2010). Fracture characterisation of short bamboo fibre reinforced polyester composites. Materials & Design, 31(9), 4147–4154.
Yushun, L., Huangying, S., Wangli, Z., & Tianshi, H. (2010). Experimental study on Bamboo-steel composite wall's seismic behavior experimental study. Engineering Mechanics, 27(1), 108–126.
Zhou, L. (2010). A new species of phellinus (hymenochaetaceae) growing on bamboo in tropical china. Mycotaxon, 114(11), 211–216.

Ahmad, M., & Kamke, F.A. (2011). Properties of parallel strand lumber from calcutta bamboo (dendrocalamus strictus). Wood Science and Technology, 45(1), 63–72.
Anwar, U.M.K., Hiziroglu, S., Hamdan, H., & Latif, M.A. (2011). Effect of outdoor exposure on some properties of resin-treated plybamboo. Industrial Crops and Products, 33(1), 140–145.
Cai, G., Wang, J., Nie, Y., Tian, X., & Zhou, X. (2011). Effects of toughening agents on the behaviors of bamboo plastic composites. Polymer Composites, 32(12).
Cheng, R., & Zhang, Q. (2011). A study on the bonding technology in sliced bamboo veneer manufacturing. Journal of Adhesion Science and Technology, 25(14), 1619–1628.
Dietsch, P. (2011). Robustness of large-span timber roof structures — structural aspects. Engineering Structures, 33(11), 3106–3112.
Guo, C., Bo, S., & Yan, X. (2011). Aseismic performance tests for a light glubam house. Journal of Vibration and Shock.
Hermawan, A., Ohuchi, T., & Fujimoto, N. (2011). Manufacture of three-layer wood–porcelain stone composite board reinforced with bamboo fiber. Materials & Design, 32(4), 2485–2489.
Kim, Y.H., Yoon, S.W., Yang, D.H., An, S.J., Kim, D.W., & Moon, K.M., et al. (2011). A study on the mechanical properties of bamboo fiber reinforced composites in comparison with matrices. Advanced Science Letters, 4(4), 1574–1577.
Li, H., & Shen, S. (2011). The mechanical properties of bamboo and vascular bundles. Journal of Materials Research, 26(21), 2749–2756.
Li, X., Wu, Y., & Zheng, X. (2011). Effect of nano anhydrous magnesium carbonateon fire-retardant performance of polylactic acid/bamboo fibers composites. Journal of Nanoscience and Nanotechnology, 11(12), 10620–10623.
Ma, H., & Whan Joo, C. (2011). Influence of surface treatments on structural and mechanical properties of bamboo fiber-reinforced poly(lactic acid) biocomposites. Journal of Composite Materials, 45(23), 2455–2463.
Mahdavi, M., Clouston, P.L., & Arwade, S.R. (2011). Development of Laminated Bamboo Lumber: Review of Processing, Performance, and Economical Considerations. Journal Of Materials In Civil Engineering, 23(7), 1036–1042.

Malanit, P., Barbu, M.C., & Arno Frühwald. (2011). Physical and mechanical properties of oriented strand lumber made from an asian bamboo (dendrocalamus asperbacker). European Journal of Wood & Wood Products, 69(1), 27–36.
Munch-Andersen, J., & Dietsch, P. (2011). Robustness of large-span timber roof structures — two examples. Engineering Structures, 33(11), 3113–3117.
Nakajima, M., Kojiro, K., Sugimoto, H., Miki, T., & Kanayama, K. (2011). Studies on bamboo for sustainable and advanced utilization. Energy, 36(4), 2049–2054.
Osorio, L., Trujillo, E., Van Vuure, A.W., & Verpoest, I. (2011). Morphological aspects and mechanical properties of single bamboo fibers and flexural characterization of bamboo/epoxy composites. Journal of Reinforced Plastics and Composites, 30(5), 396–408.
Prasad, A.V.R., & Rao, K.M. (2011). Mechanical properties of natural fibre reinforced polyester composites: jowar, sisal and bamboo. Materials & Design, 32(8–9), 4658–4663.
Richards, G. (2011). Bamboo back in the frame. Materials World, 19(11), 12–12.
Shen, H., Li, Y., Zhang, Z., & Zhang, W. (2011). Research advances of bamboo-steel composite structural members. Journal of Computational & Theoretical Nanoscience, 4(8), 2963–2967.
Sujito, & Takagi, H. (2011). Flexural strength and impact energy of microfibril bamboo fiber reinforced environment-friendly composites based on poly-lactic acid resin. International Journal of Modern Physics B, 25(31), 4195–4198.
Sun, Z.Y., Tang, Y.Q., Iwanaga, T., Sho, T., & Kida, K. (2011). Production of fuel ethanol from bamboo by concentrated sulfuric acid hydrolysis followed by continuous ethanol fermentation. Bioresource Technology, 102(23), 10929–10935.
Tan, T., Rahbar, N., Allameh, S.M., Kwofie, S., Dissmore, D., & Ghavami, K., et al. (2011). Mechanical properties of functionally graded hierarchical bamboo structures. Acta Biomaterialia, 7(10), 3796–3803.
Xu, X.S., Zhang, G., Zeng, Q.C., & Chu, H.J. (2011). Bamboo Node-Type Local Buckling of Cylindrical Shells Under Axial Impact. Advances In Vibration Engineering, 10(1), 41–52.
Yu, D., Tan, H., & Ruan, Y. (2011). A future bamboo-structure residential building prototype in china: life cycle assessment of energy use and carbon emission. Energy & Buildings, 43(10), 2638–2646.
Zhou, Q., & Xiao, Y. (2011). Flexural Behavior of FRP Reinforced Glubam Beams. Advances in FRP Composites in Civil Engineering. Springer Berlin Heidelberg.

Cheng, W., Fu, W., Zhang, Z., & Zhang, M. (2012). The Bionic Lightweight Design of the Mid-rail Box Girder Based on the Bamboo Structure. Przeglad Elektrotechniczny, 88(9B), 113–117.
Costa, S.G., Morato, E.F., & Salimon, C.I. (2012). Bamboo density and populatiuon structure of two pioneer tree species in secondary forest of different ages in a remnant forest, acre state. Scientia Forestalis/forest Sciences, 40(95), 363–374.
Feng, P., Zhang, Y., Bai, Y., & Ye, L. (2012). Combination of bamboo filling and FRP wrapping to strengthen steel members in compression. Journal of Composites for Construction, 17(3), 347–356.
Huda, S., Reddy, N., & Yang, Y. (2012). Ultra-light-weight composites from bamboo strips and polypropylene web with exceptional flexural properties. Composites Part B (Engineering), 43(3), 1658–1664.
José Jaime García, Rangel, C., & Ghavami, K. (2012). Experiments with rings to determine the anisotropic elastic constants of bamboo. Construction & Building Materials, 31(none), 52–57.
Josué Mena, Vera, S., Correal, J.F., & Lopez, M. (2012). Assessment of fire reaction and fire resistance of guadua angustifolia kunth bamboo. Construction & Building Materials, 27(1), 60–65.
Khalil, H.P.S.A., Bhat, I.U.H., Jawaid, M., Zaidon, A., Hermawan, D., & Hadi, Y.S. (2012). Bamboo fibre reinforced biocomposites: a review. Materials & Design, 42(none), 353–368.
Lee, C.H., Chung, M.J., Lin, C.H., & Yang, T.H. (2012). Effects of layered structure on the physical and mechanical properties of laminated moso bamboo (phyllosachys edulis) flooring. Construction & Building Materials, 28(1), 31–35.
Li, Y.J., Shen, Y.C., Wang, S.Q., Du, C.G., Wu, Y., & Hu, G.Y. (2012). A Dry-Wet Process to Manufacture Sliced Bamboo Veneer. Forest Products Journal, 62(5), 395–399.
Liu, D., Song, J., Anderson, D.P., Chang, P.R., & Hua, Y. (2012). Bamboo fiber and its reinforced composites: structure and properties. Cellulose, 19(5), 1449–1480.
Liu, K., Takagi, H., Osugi, R., & Yang, Z. (2012). Effect of physicochemical structure of natural fiber on transverse thermal conductivity of unidirectional abaca/bamboo fiber composites. Composites Part A (Applied Science and Manufacturing), 43(8).
Luca, V.D., & Marano, C. (2012). Prestressed glulam timbers reinforced with steel bars. Construction & Building Materials, 30(none), 206–217.
Mahdavi, M., Clouston, P.L., & Arwade, S.R. (2012). A low-technology approach toward fabrication of laminated bamboo lumber. Construction & Building Materials, 29(none), 257–262.

Pang, W., & Hassanzadeh Shirazi, S.M. (2012). Corotational model for cyclic analysis of light-frame wood shear walls and diaphragms. Journal of Structural Engineering, 139(8), 1303–1317.
Pei, S., van de Lindt, J.W., Wehbe, N., & Liu, H. (2012). Experimental study of collapse limits for wood frame shear walls. Journal of Structural Engineering, 139(9), 1489–1497.
Porras, A., & Maranon, A. (2012). Development and characterization of a laminate composite material from polylactic acid (pla) and woven bamboo fabric. Composites Part B: Engineering, 43(7), 0–0.
Ramirez, F., Correal, J.F., Yamin, L.E., Atoche, J.C., & Piscal, C.M. (2012). Dowel-Bearing Strength Behavior of Glued Laminated Guadua Bamboo. Journal Of Materials In Civil Engineering, 24(11), 1378–1387.
Sinha, A., & Clauson, M. (2012). Properties of bamboo-wood hybrid glulam beams. Forest Products Journal, 62(7/8), 541–544.
Tan, Z., Xiang, J., Su, S., Zeng, H., Zhou, C., & Sun, L., et al. (2012). Enhanced capture of elemental mercury by bamboo-based sorbents. Journal of hazardous materials, 239–240(none).
Varela, S., Correal, J., Yamin, L., & Ramirez, F. (2012). Cyclic performance of glued laminated Guadua bamboo-sheathed shear walls. Journal of structural engineering, 139(11), 2028–2037.
Verma, C.S., & Chariar, V.M. (2012). Development of layered laminate bamboo composite and their mechanical properties. Composites Part B: Engineering, 43(3), 1063–1069.
Visakh, P.M., Thomas, S., Oksman, K., & Mathew, A.P. (2012). Crosslinked natural rubber nanocomposites reinforced with cellulose whiskers isolated from bamboo waste: processing and mechanical/thermal properties. Composites Part A (Applied Science and Manufacturing), 43(4), 0–741.
Xiao, Y., & Ma, J. (2012). Fire simulation test and analysis of laminated bamboo frame building. Construction and Building Materials, 34(none), 257–266.
Xiao, Y., & Yang, R.Z. (2012). Discussion of "Development of Laminated Bamboo Lumber: Review of Processing, Performance, and Economical Considerations" by M. Mahdavi, P.L. Clouston, and S.R. Arwade. Journal Of Materials In Civil Engineering, 24(11), 1429–.
Yeh, M.C., & Lin, Y.L. (2012). Finger joint performance of structural laminated bamboo member. Journal of Wood Science, 58(2), 120–127.
Zhao, W.F., Zhou, J., & Bu, G.B. (2012). Application technology of bamboo reinforced concrete in building structure. Applied Mechanics and Materials, 195–196, 297–302.
Zhou, A., Huang, D., Li, H., & Su, Y. (2012). Hybrid approach to determine the mechanical parameters of fibers and matrixes of bamboo. Construction & Building Materials, 35(none), 191–196.
Brito, B.S.L., Pereira, F.V., Putaux, J.L., & Jean, B. (2012). Preparation, morphology and structure of cellulose nanocrystals from bamboo fibers. Cellulose, 19(5), 1527–1536.

Abdou, A., & Budaiwi, I. (2013). The variation of thermal conductivity of fibrous insulation materials under different levels of moisture content. Construction and Building materials, 43, 533–544.
Awad, Z.K., Aravinthan, T., Zhuge, Y., & Manalo, A. (2013). Geometry and restraint effects on the bending behaviour of the glass fibre reinforced polymer sandwich slabs under point load. Materials & Design, 45, 125–134.
Borri, A., Corradi, M., & Speranzini, E. (2013). Reinforcement of wood with natural fibers. Composites Part B: Engineering, 53, 1–8.
Chen, F., Jiang, Z., Wang, G., Shi, S.Q., & Liu, X.E. (2013). Bamboo bundle corrugated laminated composites (BCLC). Part I. Three-dimensional stability in response to corrugating effect. The Journal of Adhesion, 89(3), 225–238.
Cheng, D., Jiang, S., & Zhang, Q. (2013). Mould resistance of Moso bamboo treated by two step heat treatment with different aqueous solutions. European Journal of Wood and Wood Products, 71(1), 143–145.
Dongsheng, H., Aiping, Z., & Yuling, B. (2013). Experimental and analytical study on the nonlinear bending of parallel strand bamboo beams. Construction and Building Materials, 44, 585–592.
Elizabeth, S., & Datta, A.K. (2013). On the seismic performance of bamboo structure. Bulletin of Earthquake Engineering, 1–18.
Guan, M., Zhou, M., & Yong, C. (2013). Antimold effect of ultrasonic treatment on Chinese moso bamboo. Forest products journal, 63(7), 288–291.
Higuchi, I. (2013). The Stress and Strength Evaluations of Single-lap Adhesive Joints of Bamboo Plate Adherends under Static Bending Moments. Mokuzai Gakkaishi, 59(6), 383–390.
Hu, J.B., & Pizzi, A. (2013). Wood–bamboo–wood laminated composite lumber jointed by linear vibration–friction welding. European Journal of Wood and Wood Products, 71(5), 683–686.
Kumar, P.S., Kanwat, M., & Choudhary, V.K. (2013). Mathematical modeling and thin-layer drying kinetics of bamboo slices on convective tray drying at varying temperature. Journal of food processing and preservation, 37(5), 914–923.
Li, H.T., Zhang, Q.S., Huang, D.S., & Deeks, A.J. (2013). Compressive performance of laminated bamboo. Composites Part B: Engineering, 54, 319–328.

Liu, J., Zhang, H., Chrusciel, L., Na, B., & Lu, X. (2013). Study on a bamboo stressed flattening process. European Journal of Wood and Wood Products, 71(3), 291–296.
Lu, T., Jiang, M., Jiang, Z., Hui, D., Wang, Z., & Zhou, Z. (2013). Effect of surface modification of bamboo cellulose fibers on mechanical properties of cellulose/epoxy composites. Composites Part B: Engineering, 51, 28–34.
Marinho, N.P., Nascimento, E.M.D., Nisgoski, S., & Valarelli, I.D.D. (2013). Some physical and mechanical properties of medium-density fiberboard made from giant bamboo. Materials Research, 16(6), 1398–1404.
Raftery, G.M., & Harte, A.M. (2013). Nonlinear numerical modelling of FRP reinforced glued laminated timber. Composites Part B: Engineering, 52, 40–50.
Shao, J., Wang, F., Li, L., & Zhang, J. (2013). Scaling analysis of the tensile strength of bamboo fibers using Weibull statistics. Advances in Materials Science and Engineering, 2013.
Sieder, M., Rein, A., & Seise, N. (2013). Vom Halm zum Tragwerk: Bambuspavillon EXPO Shanghai 2010–Eine deutsch-chinesische Kooperation. Bautechnik, 90(12), 816–821.
Thite, A.N., Gerguri, S., Coleman, F., Doody, M., & Fisher, N. (2013). Development of an experimental methodology to evaluate the influence of a bamboo frame on the bicycle ride comfort. Vehicle System Dynamics, 51(9), 1287–1304.
Tsuchiya, Y., & Higuchi, I. (2013). Stress and Strength of a Laminated Composite Plate Subjected to Static Bending Moments. Mokuzai Gakkaishi, 59(3), 128–137.
Verma, C.S., & Chariar, V.M. (2013). Stiffness and strength analysis of four layered laminate bamboo composite at macroscopic scale. Composites Part B: Engineering, 45(1), 369–376.
Wang, F., Shao, Z., & Wu, Y. (2013). Mode II interlaminar fracture properties of Moso bamboo. Composites Part B: Engineering, 44(1), 242–247.
Xiao, Y., Yang, R.Z., & Shan, B. . (2013). Production, environmental impact and mechanical properties of glubam. Construction and Building Materials, 44, 765–773.
Xing, D., Chen, W., Xing, D., & Yang, T. (2013). Lightweight design for thin-walled cylindrical shell based on action mechanism of bamboo node. Journal of Mechanical Design, 135(1), 014501.
Zhang, H., Liu, J., Wang, Z., & Lu, X. (2013). Mechanical and thermal properties of small diameter original bamboo reinforced extruded particleboard. Materials Letters, 100, 204–206.
Zhang, Y.M., Yu, Y.L., & Yu, W.J. (2013). Effect of thermal treatment on the physical and mechanical properties of Phyllostachys pubescen bamboo. European Journal of Wood and Wood Products, 71(1), 61–67.

Agarwal, A., Nanda, B., & Maity, D. (2014). Experimental investigation on chemically treated bamboo reinforced concrete beams and columns. Construction and Building Materials, 71, 610–617.
Aiping, Z., & Yuling, B. (2014). Experimental study on the flexural performance of parallel strand bamboo beams. The Scientific World Journal, 2014, 1–6.
Chen, F., Deng, J., Cheng, H., Li, H., & Shi, S.Q. (2014). Impact properties of bamboo bundle laminated veneer lumber by preprocessing densification technology. Journal of Wood Science, 60(6).
Correal, J.F., Echeverry, J.S., Ramírez, Fernando, & Yamín, Luis E. (2014). Experimental evaluation of physical and mechanical properties of glued laminated guadua angustifolia kunth. Construction and Building Materials, 73, 105–112.
Desalegn, G., & Tadesse, W. (2014). Resource potential of bamboo, challenges and future directions towards sustainable management and utilization in ethiopia. Forest Systems.
Dixon, P.G., & Gibson, L.J. (2014). The structure and mechanics of moso bamboo material. Journal of the Royal Society Interface, 11(99).
Fu, X.F., Yong, C., & Guan, M.J. (2014). The aging properties of bamboo-poplar composite oriented strand board with different hybrid ratios. Applied Mechanics & Materials, 599-601, 140–143.
Gatoo, A., Sharma, B., Bock, M., Mulligan, H., & Ramage, M.H. (2014). Sustainable structures: bamboo standards and building codes. Proceedings Of the Institution Of Civil Engineers-Engineering Sustainability, 167(5), 189–196.
Gottron, J., Harries, K.A., & Xu, Q. (2014). Creep behaviour of bamboo. Construction and Building Materials, 66, 79–88.
Hebel, D.E., Javadian, A., Heisel, F., Schlesier, K., Griebel, D., & Wielopolski, M. (2014). Process-controlled optimization of the tensile strength of bamboo fiber composites for structural applications. Composites Part B: Engineering, 67, 125–131.
Luna, P., & Takeuchi, C. (2014). Home for elderly people built by the community with structural elements of laminated bamboo guadua in a rural area of colombia. Key Engineering Materials, 600, 773–782.
Luna, P., Takeuchi, C., & Cordón, Edwar. (2014). Mechanical behavior of glued laminated pressed bamboo guadua using different adhesives and environmental conditions. Key Engineering Materials, 600, 57–68.

Moroz, J.G., Lissel, S.L., & Hagel, M.D. (2014). Performance of bamboo reinforced concrete masonry shear walls. Construction and Building Materials, 61, 125–137.
Mujiman, Priyosulistyo, H., Sulistyo, D., & Prayitno, T.A. (2014). Influence of shape and dimensions of lamina on shear and bending strength of vertically glue laminated bamboo beam. Procedia Engineering, 95, 22–30.
Qi, J.Q., Xie, J.L., Huang, X.Y., Yu, W.J., & Chen, S.M. (2014). Influence of characteristic inhomogeneity of bamboo culm on mechanical properties of bamboo plywood: effect of culm height. Journal of Wood Science, 60(6), 396–402.
Rassiah, K., Ahmad, M.M.H.M., & Ali, A. (2014). Mechanical properties of laminated bamboo strips from gigantochloa scortechinii/polyester composites. Materials & Design, 57(5), 551–559.
Ren, D., Yu, Z., Li, W., Wang, H., & Yu, Y. (2014). The effect of ages on the tensile mechanical properties of elementary fibers extracted from two sympodial bamboo species. Industrial Crops and Products, 62, 94–99.
Rosa, R.A., Paes, J.B., Segundinho, P.G.D.A., Vidaurre, G.B., & Gonçalves, F.G. (2014). Effects of preservative treatment and the adhesive on mechanical characteristics of laminated lumber of two bamboo species. Scientia Forestalis/forest Sciences, 42(103), 451–462.
Sinha, A., Way, D., & Mlasko, S. (2014). Structural Performance of Glued Laminated Bamboo Beams. Journal Of Structural Engineering, 140(1).
Verma, C.S., Sharma, N.K., Chariar, V.M., Maheshwari, S., & Hada, M.K. (2014). Comparative study of mechanical properties of bamboo laminae and their laminates with woods and wood based composites. Composites Part B: Engineering, 60, 523–530.
Wang, H., An, X., Li, W., Wang, H., & Yu, Y. (2014). Variation of mechanical properties of single bamboo fibers (Dendrocalamus latiflorus Munro) with respect to age and location in culms. Holzforschung, 68(3), 291–297.
Wu, W. (2014). Experimental Analysis of Bending Resistance of Bamboo Composite I-Shaped Beam. Journal Of Bridge Engineering, 19(4).
Wu, W.Q., Wu, Z.Z., Ma, X.Y., Chen, S., & Wei, H.W. (2014). Test on bending resistance properties of bamboo-based composite i-shaped beam. Zhongguo Gonglu Xuebao/China Journal of Highway and Transport, 27(4), 69–78.
Wu, X.F., Xu, J.Y., Hao, J.X., Liao, R., & Zhong, Z. (2014). Three-point flexural normal stress analysis of wood-bamboo sandwich composite. Applied Mechanics and Materials, 672–674, 1894–1898.
Xiao, S., Lin, H., Shi, S.Q., & Cai, L. (2014). Optimum processing parameters for wood-bamboo hybrid composite sleepers. Journal Of Reinforced Plastics And Composites, 33(21), 2010–2018.
Xiao, Y., Chen, G., & Feng, L. (2014). Experimental studies on roof trusses made of glubam. Materials & Structures, 47(11), 1879–1890.
Xiao, Y., Shan, B., Yang, R.Z., Li, Z., & Chen, J. (2014). Glue laminated bamboo (glubam) for structural applications. RILEM Bookseries, 9(1), 589–601.
Xiao, Y., Li, L., & Yang, R.Z. (2014). Long-Term Loading Behavior of a Full-Scale Glubam Bridge Model. Journal Of Bridge Engineering, 19(9).
Yang, F., Fei, B., Wu, Z., Peng, L., & Yu, Y. (2014). Selected Properties of Corrugated Particleboards Made from Bamboo Waste (Phyllostachys edulis) Laminated with Medium-Density Fiberboard Panels. Bioresources, 9(1), 1085–1096.
Yang, R.Z., Xiao, Y., & Lam, F. (2014). Failure analysis of typical glubam with bidirectional fibers by off-axis tension tests. Construction and Building Materials, 58, 9–15.
Young, W.B., & Tsao, Y.C. (2014). The mechanical and fire safety properties of bamboo fiber reinforced polylactide biocomposites fabricated by injection molding. Journal of Composite Materials, 49(22).
Yu, Y., Wang, H., Lu, F., Tian, G., & Lin, J. (2014). Bamboo fibers for composite applications: a mechanical and morphological investigation. Journal of Materials Science, 49(6), 2559–2566.
Zea Escamilla, E., & Habert, G. (2014). Environmental impacts of bamboo-based construction materials representing global production diversity. Journal of Cleaner Production, 69, 117–127.
Zhao, W., Zhang, W., Zhou, J., Cao, Y., & Long, Z. (2014). Axial compression behavior of square thin-walled steel tube-laminated bamboo composite hollow columns. Transactions of the Chinese Society of Agricultural Engineering, 30(6), 37–45.
Zheng, Y., Jiang, Z., Sun, Z., & Ren, H. 2014). Effect of microwave-assisted curing on bamboo glue strength: bonded by thermosetting phenolic resin. Construction and Building Materials, 68, 320–325.

Armandei, M., Darwish, I.F., & Ghavami, K. (2015). Experimental study on variation of mechanical properties of a cantilever beam of bamboo. Construction and Building Materials, 101, 784–790.

Chen, G., Li, H.T., Zhou, T., Li, C.L., Song, Y.Q., & Xu, R. (2015). Experimental evaluation on mechanical performance of osb webbed parallel strand bamboo i-joist with holes in the web. Construction and Building Materials, 101, 91–98.
Chen, J.H., Wang, K., Xu, F., & Sun, R.C. (2015). Effect of hemicellulose removal on the structural and mechanical properties of regenerated fibers from bamboo. Cellulose, 22(1), 63–72.
Cheng, H.T., Gao, J., Wang, G., Shi, S.Q., Zhang, S.B., & Cai, L.P. (2015). Enhancement of mechanical properties of composites made of calcium carbonate modified bamboo fibers and polypropylene. Holzforschung, 69(2), 215–221.
Deng, J., Li, H., Wang, G., Chen, F., & Zhang, W. (2015). Effect of removing extent of bamboo green on physical and mechanical properties of laminated bamboo-bundle veneer lumber (blvl). Holz als Roh- und Werkstoff, 73(4), 499–506.
Dixon, P.G., Ahvenainen, P., Aijazi, A.N., Chen, S.H., Lin, S., Augusciak, P.K., Gibson, L.J. (2015). Comparison of the structure and flexural properties of Moso, Guadua and Tre Gai bamboo. Construction And Building Materials, 90, 11–17.
Fang, H., Sun, H.M., Liu, W.Q., Wang, L., Bai, Y., & Hui, D. (2015). Mechanical performance of innovative GFRP-bamboo-wood sandwich beams: Experimental and modelling investigation. Composites Part B-Engineering, 79, 182–196.
Fuentes, C.A., Brughmans, G., Tran, L.Q.N., Dupont-Gillain, C., Verpoest, I., & Van Vuure, A.W. (2015). Mechanical behaviour and practical adhesion at a bamboo composite interface: physical adhesion and mechanical interlocking. Composites Science and Technology, 109, 40–47.
He, M., Li, Z., Sun, Y., & Ma, R. (2015). Experimental investigations on mechanical properties and column buckling behavior of structural bamboo. The Structural Design of Tall and Special Buildings, 24(7), 491–503.
Hegde, A., & Sitharam, T.G. (2015). Use of bamboo in soft-ground engineering and its performance comparison with geosynthetics: experimental studies. Journal of Materials in Civil Engineering, 27(9).
Huang, D., Bian, Y., Zhou, A., & Sheng, B. (2015). Experimental study on stress–strain relationships and failure mechanisms of parallel strand bamboo made from phyllostachys. Construction and Building Materials, 77, 130–138.
Huang, X.D., Shupe, T.F., & Hse, C.Y. (2015). Study of moso bamboo's permeability and mechanical properties. Emerging Materials Research, 4(1), 130–138.
Korde, C., West, R., Gupta, A., & Puttagunta, S. (2015). Laterally Restrained Bamboo Concrete Composite Arch under Uniformly Distributed Loading. Journal Of Structural Engineering, 141(3), 11.
Li, H.T., Su, J.W., Zhang, Q.S., Deeks, A.J., & Hui, D. (2015). Mechanical performance of laminated bamboo column under axial compression. Composites Part B-Engineering, 79, 374–382.
Li, T., Cheng, D.L., Walinder, M.E.P., & Zhou, D.G. (2015). Wettability of oil heat-treated bamboo and bonding strength of laminated bamboo board. Industrial Crops And Products, 69, 15–20.
Li, Y., Shan, W., Shen, H., Zhang, Z.W., & Liu, J. (2015). Bending resistance of i-section bamboo–steel composite beams utilizing adhesive bonding. Thin-Walled Structures, 89, 17–24.
Li, Z., Xiao, Y., Wang, R., & Monti, G. (2015). Studies of Nail Connectors Used in Wood Frame Shear Walls with Ply-Bamboo Sheathing Panels. Journal Of Materials In Civil Engineering, 27(7), 13.
Ma, X.X., Jiang, Z.H., Tong, L., Wang, G., & Cheng, H.T. (2015). Development of Creep Models for Glued Laminated Bamboo Using the Time-Temperature Superposition Principle. Wood And Fiber Science, 47(2), 141–146.
Manalo, A.C., Wani, E., Zukarnain, N.A., Karunasena, W., & Lau, K.T. (2015). Effects of alkali treatment and elevated temperature on the mechanical properties of bamboo fibre-polyester composites. Composites Part B Engineering, 80(2), 73–83.
Polito, N., Burdiak, T., Siniawski, M.T., & Eberts, W. (2015). Mechanical characterization of bamboo and glass fiber biocomposite laminates. Journal of Renewable Materials, 3(4).
Potwar, K., Ackerman, J., & Seipel, J. (2015). Design of Compliant Bamboo Poles for Carrying Loads. Journal Of Mechanical Design, 137(1), 14. doi: 10.1115/1.4028757.
Qi, J., Xie, J., Yu, W., & Chen, S. (2015). Effects of characteristic inhomogeneity of bamboo culm nodes on mechanical properties of bamboo fiber reinforced composite. Journal of Forestry Research, 26(4), 1057–1060.
Richard, M.J., & Harries, K.A. (2015). On inherent bending in tension tests of bamboo. Wood Science and Technology, 49(1), 99–119.
Shang, L., Sun, Z., Liu, Xing'e, & Jiang, Z. (2015). A novel method for measuring mechanical properties of vascular bundles in moso bamboo. Journal of Wood Science, 61(6), 562–568.
Shangguan, W., Zhong, Y., Xing, X., Zhao, R., & Ren, H. (2015). Strength models of bamboo scrimber for compressive properties. Journal of Wood Science, 61(2), 120–127.

Sharma, B., Gatóo, Ana, & Ramage, M.H. (2015). Effect of processing methods on the mechanical properties of engineered bamboo. Construction and Building Materials, 83, 95–101.
Sharma, B., Gatóo, Ana, Bock, M., & Ramage, M. (2015). Engineered bamboo for structural applications. Construction and Building Materials, 81, 66–73.
Song, W., Zhao, F., Yu, X.F., Wang, C.C., Wei, W.B., & Zhang, S.B. (2015). Interfacial Characterization and Optimal Preparation of Novel Bamboo Plastic Composite Engineering Materials. Bioresources, 10(3), 5049–5070.
Teixeira, D.E., & Bastos, R.P. (2015). Characterization of glued laminated panels produced with strips of bamboo (guadua magna) native from the brazilian cerrado. Cerne,21(4), 595–600.
Tsuchiya, Y., & Higuchi, I. (2015). Stress Analysis and Strength of Single Strapped Adhesive Joints Made from Bamboo and Hinoki Subjected to Bending Moments. Mokuzai Gakkaishi, 61(4), 280–290.
Wang, X.H., Wu, Z.H., Wang, X.H., Song, S.S., Cao, Y.J., Jun, N., Fei, B.H. (2015). Shearing behavior of structural insulated panel wall shelled with bamboo scrimber. Wood And Fiber Science, 47(4), 336–344.
Wei, Y., Zhou, M.Q., & Chen, D.J. (2015). Flexural behaviour of glulam bamboo beams reinforced with near-surface mounted steel bars. Materials Research Innovations, 19(S1), S1-98-S1–103.
Xiao, Y., Li, Z., & Wang, R. (2015). Lateral Loading Behaviors of Lightweight Wood-Frame Shear Walls with Ply-Bamboo Sheathing Panels. Journal Of Structural Engineering, 141(3).
Yu, Y., Zhu, R., Wu, B., Hu, Yu'an, & Yu, W. (2015). Fabrication, material properties, and application of bamboo scrimber. Wood Science and Technology, 49(1), 83–98.
Yu, Y.S., Ni, C.Y., Yu, T., & Wan, H. (2015). Optimization of mechanical properties of bamboo plywood. Wood And Fiber Science, 47(1), 109–119.
Zhou, J.B., Fu, W.S., Qing, Y., Han, W., Zhao, Z.R., & Zhang, B. (2015). Fabrication and Performance of a Glue-Pressed Engineered Honeycomb Bamboo (GPEHB) Structure with Finger-jointed Ends as a Potential Substitute for Wood Lumber. Bioresources, 10(2), 3302–3313.
Ziemann, M., Chen, Y., Kraft, O., Bayerschen, E., Wulfinghoff, S., & Kirchlechner, C., et al. (2015). Deformation patterns in cross-sections of twisted bamboo-structured au microwires. Acta Materialia, 97, 216–222.
Zou, M., Wei, C.G., Li, J.Q., Xu, S.C., & Zhang, X. (2015). The energy absorption of bamboo under dynamic axial loading. Thin-Walled Structures, 95, 255–261.

Anokye, R., Bakar, E.S., Ratnasingam, J., Choo, A., Yong, C., & Bakar, N.N. (2016). The effects of nodes and resin on the mechanical properties of laminated bamboo timber produced from gigantochloa scortechinii. Construction & Building Materials, 105(2016), 285–290.
Ashaari, Z., Lee, S.H., & Zahali, M.R. (2016). Performance ofcompreglaminated bamboo/wood hybrid using phenolic-resin-treated strips as core layer. European Journal of Wood and Wood Products, 74(4), 621–624.
Askarinejad, S., Kotowski, P., Youssefian, S., & Rahbar, N. (2016). Fracture and mixed-mode resistance curve behavior of bamboo. Mechanics Research Communications, 78, 79–85.
Castanet, E., Li, Q., Dumée, Ludovic F., Garvey, C., Rajkhowa, R., & Zhang, J., et al. (2016). Structure–property relationships of elementary bamboo fibers. Cellulose, 23(6), 3521–3534.
Chen, F., Jiang, Z., Wang, G., Li, H., Simth, L.M., & Shi, S.Q. (2016). The bending properties of bamboo bundle laminated veneer lumber (blvl) double beams. Construction and Building Materials, 119, 145–151.
Chen, F.M., Wang, G., Li, X.J., Simth, L.M., & Shi, S.Q. (2016). Laminated structure design of wood-Bamboo hybrid laminated composite using finite element simulations. Journal Of Reinforced Plastics And Composites, 35(22), 1661–1670.
Chen, H., Cheng, H., Wang, G., Yu, Z., & Shi, S.Q. (2016). Erratum to: tensile properties of bamboo in different sizes. Journal of Wood Science, 62(2), 213–213.
Chen, Y.X., Zhu, S.L., Guo, Y., Liu, S.Q., Tu, D.W., & Fan, H. (2016). Investigation on withdrawal resistance of screws in reconstituted Bamboo lumber. Wood Research, 61(5), 799–810.
Dai, Z.H., Guo, W.D., Zheng, G.X., Ou, Y., & Chen, Y.J. (2016). Moso Bamboo Soil-Nailed Wall and Its 3D Nonlinear Numerical Analysis. International Journal Of Geomechanics, 16(5), 14.
Deng, J., Chen, F., Li, H., Wang, G., & Shi, S.Q. (2016). The effect of pf/pvac weight ratio and ambient temperature on moisture absorption performance of bamboo-bundle laminated veneer lumber. Polymer Composites, 37(3).
Deng, J.C., Chen, F.M., Wang, G., & Zhang, W.F. (2016). Variation of Parallel-to-Grain Compression and Shearing Properties in Moso Bamboo Culm (Phyllostachys pubescens). Bioresources, 11(1), 1784–1795.
Dixon, P.G., Semple, K.E., Kutnar, A., Kamke, F.A., Smith, G.D., & Gibson, L.J. (2016). Comparison of the flexural behavior of natural and thermo-hydro-mechanically densified moso bamboo. European Journal of Wood and Wood Products, 74(5), 633–642.
Gong, Y.C., Zhang, C.Q., Zhao, R.J., Xing, X.T., & Ren, H.Q. (2016). Experimental Study on Tensile and Compressive Strength of Bamboo Scrimber. Bioresources, 11(3), 7334–7344.

Guo, N., Chen, H.H., Zhang, P.Y., & Zuo, H.L. (2016). The research of parallel to the grain compression performance test of laminated glued bamboo-wood composites. Tehnicki Vjesnik-Technical Gazette, 23(1), 129–135.

Habibi, M.K., Tam, L.H., Lau, D., & Lu, Y. (2016). Viscoelastic damping behavior of structural bamboo material and its microstructural origins. Mechanics of Materials, S0167663616000521.

Hai-Tao, L., John, D.A., Qisheng, Z., & Gang, W. . (2016). Flexural performance of laminated bamboo lumber beams. BioResources, 11(1).

He, M.J., Zhang, J., Li, Z., & Li, M.L. (2016). Production and mechanical performance of scrimber composite manufactured from poplar wood for structural applications. Journal of Wood Science, 62, 1–12.

Huang, Z., Chen, Z., Huang, D., & Zhou, A. (2016). The ultimate load-carrying capacity and deformation of laminated bamboo hollow decks: experimental investigation and inelastic analysis. Construction and Building Materials, 117, 190–197.

Kumar, A., Vlach, Tomá, Laiblova, L., Hrouda, M., Kasal, B., & Tywoniak, J., et al. (2016). Engineered bamboo scrimber: influence of density on the mechanical and water absorption properties. Construction and Building Materials, 127, 815–827.

Latha, P.S., Rao, M.V., Kumar, V.V.K., Raghavendra, G., Ojha, S., & Inala, R. (2016). Evaluation of mechanical and tribological properties of bamboo-glass hybrid fiber reinforced polymer composite. Journal Of Industrial Textiles, 46(1), 3–18.

Li, H.D., Xian, Y., Deng, J.C., Cheng, H.T., Chen, F.M., & Wang, G. (2016). Evaluation of Water Absorption and its Influence on the Physical-Mechanical Properties of Bamboo-Bundle Laminated Veneer Lumber. Bioresources, 11(1), 1359–1368.

Li, H.T., Chen, G., Zhang, Q., Ashraf, M., Xu, B., & Li, Y. (2016). Mechanical properties of laminated bamboo lumber column under radial eccentric compression. Construction and Building Materials, 121, 644–652.

Li, H.T., Wu, G., Zhang, Q.S., & Su, J.W. (2016). Mechanical evaluation for laminated bamboo lumber along two eccentric compression directions. Journal of Wood Science, 62(6), 503–517.

Li, J.Q., Yuan, Y., & Guan, X. (2016). Assessing the Environmental Impacts of Glued-Laminated Bamboo Based on a Life Cycle Assessment. Bioresources, 11(1), 1941–1950.

Li, L., & Xiao, Y. (2016). Creep Behavior of Glubam and CFRP-Enhanced Glubam Beams. Journal Of Composites for Construction, 20(1), 11.

Li, Z., He, M., Tao, D., & Li, M. (2016). Experimental buckling performance of scrimber composite columns under axial compression. Composites Part B: Engineering, 86, 203–213.

Ni, L., Zhang, X.B., Liu, H.R., Sun, Z.J., Song, G.N., Yang, L.M., & Jiang, Z.H. (2016). Manufacture and Mechanical Properties of Glued Bamboo Laminates. Bioresources, 11(2), 4459–4471.

Richard, M.J., Kassabian, P.E., & Schulze-Ehring, H.S. (2016). Bamboo active school: structural design and material testing. Proceedings of the Institution of Civil Engineers-Structures and Buildings, 170(4), 275–283.

Rosa, R.A., Paes, J.B., Segundinho, P.G.D., Vidaurre, G.B., & de Oliveira, A.K.F. (2016). Influences of species, preservative treatment and adhesives on physical properties of laminated bamboo lumber. Ciencia Florestal, 26(3), 913–924.

Salcido, J.C., Raheem, A.A., & Ravi, S. (2016). Comparison of embodied energy and environmental impact of alternative materials used in reticulated dome construction. Building and Environment, 96, 22–34.

Sassu, M., De Falco, A., Giresini, L., & Puppio, M. (2016). Structural solutions for low-cost bamboo frames: experimental tests and constructive assessments. Materials, 9(5).

Shangguan, W.W., Gong, Y.C., Zhao, R.J., & Ren, H.Q. (2016). Effects of heat treatment on the properties of bamboo scrimber. Journal Of Wood Science, 62(5), 383–391.

Sharma, B., Bauer, H., Schickhofer, G., & Ramage, M.H. (2016). Mechanical characterisation of structural laminated bamboo. Proceedings of the Institution of Civil Engineers-Structures and Buildings, 170(4), 250–264.

Sukmawan, R., Takagi, H., & Nakagaito, A.N. (2016). Strength evaluation of cross-ply green composite laminates reinforced by bamboo fiber. Composites Part B: Engineering, 84, 9–16.

Xie, J., Qi, J., Hu, T., De Hoop, C.F., Hse, C.Y., & Shupe, T.F. (2016). Effect of fabricated density and bamboo species on physical–mechanical properties of bamboo fiber bundle reinforced composites. Journal of Materials Science, 51(16), 7480–7490.

Yan, Y., Liu, H., Zhang, X., Wu, H., & Huang, Y. (2016). The effect of depth and diameter of glued-in rods on pull-out connection strength of bamboo glulam. Journal of Wood Science, 62(1), 109–115.

Zhang, Z.W., Li, Y.S., & Liu, R. (2016). Failure behavior of adhesive bonded interface between steel and bamboo plywood. Journal of Adhesion Science and Technology, 30(19), 1–19.

Zhao, W.F., Zhou, J., & Long, Z.L. (2016). Compression tests on square, thin-walled steel tube/bamboo--plywood composite hollow columns. Science And Engineering Of Composite Materials, 23(5), 511–522.

Zhong, Y., Ren, H.Q., & Jiang, Z.H. (2016). Effects of temperature on the compressive strength parallel to the grain of bamboo scrimbe. Materials, 9(6), 436.
Zou, M., Xu, S.C., Wei, C.G., Wang, H.X., & Liu, Z.Z. (2016). A bionic method for the crashworthiness design of thin-walled structures inspired by bamboo. Thin-Walled Structures, 101, 222–230.
Zuhudi, N.Z.M., Lin, R.J.T., & Jayaraman, K. (2016). Flammability, thermal and dynamic mechanical properties of bamboo-glass hybrid composites. Journal Of Thermoplastic Composite Materials, 29(9), 1210–1228.

Bahari, S.A., Grigsby, W.J., & Krause, A. (2017). Flexural properties of PVC/bamboo composites under static and dynamic-thermal conditions: effects of composition and water absorption. International Journal of Polymer Science, 2017.
Carrasco, E.V.M., Smits, M.A., & Mantilla, J.N.R. (2017). Shear strength of bamboo-bamboo connection: gluing pressure Influence. Matéria (Rio de Janeiro), 22.
Chen, F., Deng, J., Li, X., Wang, G., Smith, L.M., & Shi, S.Q. (2017). Effect of laminated structure design on the mechanical properties of bamboo-wood hybrid laminated veneer lumber. European Journal of Wood and Wood Products, 75(3), 439–448.
Chithambaram, S.J., & Kumar, S. (2017). Flexural behaviour of bamboo based ferrocement slab panels with flyash. Construction and Building Materials, 134, 641–648.
Dixon, P.G., Muth, J.T., Xiao, X., Skylar-Scott, M.A., Lewis, J.A., & Gibson, L.J. (2017). 3d printed structures for modeling the young's modulus of bamboo parenchyma. Acta Biomaterialia, S1742706117308024.
Dixon, P.G., Malek, S., Semple, K.E., Zhang, P.K., Smith, G.D., & Gibson, L.J. (2017). Multiscale modelling of moso bamboo oriented strand board.
Fu, Y., Fang, H., & Dai, F. (2017). Study on the properties of the recombinant bamboo by finite element method. Composites Part B: Engineering, 115, 151–159.
Gao, W.C., & Xiao, Y. (2017). Seismic behavior of cold-formed steel frame shear walls sheathed with ply-bamboo panels. Journal of Constructional Steel Research, 132, 217–229.
Harries, K.A., Bumstead, J., Richard, M., & Trujillo, D. (2017). Geometric and material effects on bamboo buckling behaviour. Proceedings of the Institution of Civil Engineers-Structures and Buildings, 170(4).
Huang, Z., Sun, Y., & Musso, F. (2017). Experimental study on bamboo hygrothermal properties and the impact of bamboo-based panel process. Construction and Building Materials, 155, 1112–1125.
Jakovljević, S., Lisjak, D., Alar, Ž., & Penava, F. (2017). The influence of humidity on mechanical properties of bamboo for bicycles. Construction and Building Materials, 150, 35–48.
Khan, Z., Yousif, B.F., & Islam, M. (2017). Fracture behaviour of bamboo fiber reinforced epoxy composites. Composites Part B: Engineering, 116, 186–199.
Li, H., Chen, F., Xian, Y., Deng, J., Wang, G., & Cheng, H. (2017). An empirical model for predicting the mechanical properties degradation of bamboo bundle laminated veneer lumber (BLVL) by hygrothermal aging treatment. European Journal of Wood and Wood Products, 75(4), 553–560.
Li, W.T., Long, Y.L., Huang, J., & Lin, Y. (2017). Axial load behavior of structural bamboo filled with concrete and cement mortar. Construction and Building Materials, 148, 273–287.
Li, Y., Yao, J., Li, R., Zhang, Z., & Zhang, J. (2017). Thermal and energy performance of a steel-bamboo composite wall structure. Energy and Buildings, 156, 225–237.
Liu, H., Jiang, Z., Sun, Z., Yan, Y., Cai, Z., & Zhang, X. (2017). Impact performance of two bamboo-based laminated composites. European Journal of Wood and Wood Products, 75(5), 711–718.
Lorenzo, R., Lee, C., Oliva-Salinas, J.G., & Ontiveros-Hernandez, M.J. (2017). BIM Bamboo: a digital design framework for bamboo culms. Proceedings of the Institution of Civil Engineers-Structures and Buildings, 170(4), 295–302.
Ma, X., Smith, L.M., Wang, G., Jiang, Z., & Fei, B. (2017). Mechano-sorptive creep mechanism of wood in compression and bending. Wood and Fiber Science, 49(3), 1–8.
Mannan, S., Paul Knox, J., & Basu, S. (2017). Correlations between axial stiffness and microstructure of a species of bamboo. Royal Society open science, 4(1), 160412.
Moran, R., Ghavami, K., & García, J.J. (2017). A new method to measure the axial and shear moduli of bamboo. Proceedings of the Institution of Civil Engineers-Structures and Buildings, 170(4), 303–310.
Moran, R., Webb, K., Harries, K., & García, J.J. (2017). Edge bearing tests to assess the influence of radial gradation on the transverse behavior of bamboo. Construction and Building Materials, 131, 574–584.
Murugan, R., Karthik, T., Dasardan, B.S., Subramanian, V., & Shanmugavadivu, K. (2017). Effect of lateral crushing on tensile property of bamboo, modal and tencel fibres.
Nugroho, N., & Bahtiar, E.T. (2017). Structural grading of Gigantochloa apus bamboo based on its flexural properties. Construction and Building Materials, 157, 1173–1189.

Ospina-Borras, J., Benitez-Restrepo, H., & Florez-Ospina, J. (2017). Non-Destructive Infrared Evaluation of Thermo-Physical Parameters in Bamboo Specimens. Applied Sciences, 7(12), 1253.
Paraskeva, T.S., Grigoropoulos, G., & Dimitrakopoulos, E.G. (2017). Design and experimental verification of easily constructible bamboo footbridges for rural areas. Engineering structures, 143, 540–548.
Puri, V., Chakrabortty, P., Anand, S., & Majumdar, S. (2017). Bamboo reinforced prefabricated wall panels for low cost housing. Journal of Building Engineering, 9, 52–59.
Ramage, M.H., Sharma, B., Shah, D.U., & Reynolds, T.P. (2017). Thermal relaxation of laminated bamboo for folded shells. Materials & Design, 132, 582–589.
Rassiah, K., Ahmad, M.M.H.M., Ali, A., Abdullah, A.H., & Nagapan, S. (2017). Mechanical properties of layered laminated woven bamboo gigantochloa scortechinii/epoxy composites. Journal of Polymers and the Environment (8), 1–15.
Ribeiro, R.A.S., Ribeiro, M.G.S., & Miranda, I.P. (2017). Bending strength and nondestructive evaluation of structural bamboo. Construction and building materials, 146, 38–42.
Sato, M., Inoue, A., & Shima, H. (2017). Bamboo-inspired optimal design for functionally graded hollow cylinders. PloS one, 12(5), e0175029.
Singh, T.J., & Samanta, S. (2017). Effect of stacking sequence on mechanical strength of bamboo/Kevlar K29 inter-ply laminated hybrid composite.
Sinha, A.K., Narang, H.K., & Bhattacharya, S. (2017). Mechanical properties of natural fibre polymer composites. Journal of Polymer Engineering, 37(9), 879–895.
Song, J., Gao, L., & Lu, Y. (2017). In Situ mechanical characterization of structural bamboo materials under flexural bending. Experimental Techniques, 41(6), 565–575.
Song, J., Surjadi, J.U., Hu, D., & Lu, Y. (2017). Fatigue characterization of structural bamboo materials under flexural bending. International Journal of Fatigue, 100, 126–135.
Srivaro, Suthon. (2017). Potential of three sympodial bamboo species naturally growing in thailand for structural application. European Journal of Wood and Wood Products.
Trujillo, D., Jangra, S., & Gibson, J.M. (2017). Flexural properties as a basis for bamboo strength grading. Proceedings of the Institution of Civil Engineers–Structures and Buildings, 170(4), 284–294.
Wang, R., Xiao, Y., & Li, Z. (2017). Lateral loading performance of lightweight glubam shear walls. Journal of Structural Engineering, 143(6), 04017020.
Wei, Y., Ji, X., Duan, M., & Li, G. (2017). Flexural performance of bamboo scrimber beams strengthened with fiber-reinforced polymer. Construction and Building Materials, 142, 66–82.
Wong, K.J., Low, K.O., & Israr, H.A. (2017). Impact resistance of short bamboo fibre reinforced polyester concretes. Proceedings of the Institution of Mechanical Engineers, Part L: Journal of Materials: Design and Applications, 231(8), 683–692.
Xiao, Y., Wu, Y., Li, J., & Yang, R.Z. (2017). An experimental study on shear strength of glubam. Construction and Building Materials, 150, 490–500.
Xie, J., Qi, J., Hu, T., Xiao, H., Chen, Y., Cornelis, F., & Huang, X. (2017). Anatomical characteristics and physical–mechanical properties of Neosinocalamus affinis from Southwest China. European journal of wood and wood products, 75(4), 659–662.
Xu, M., Cui, Z., Chen, Z., & Xiang, J. (2017). Experimental study on compressive and tensile properties of a bamboo scrimber at elevated temperatures. Construction and Building Materials, 151, 732–741.
Xu, Q., Chen, L., Harries, K.A., & Li, X. (2017). Combustion performance of engineered bamboo from cone calorimeter tests. European Journal of Wood and Wood Products, 75(2), 161–173.
Yang, F., & Hai, F. (2017). Investigation of performance of recombinant bamboo chair through finite element technology. Wood Research, 62(6), 995–1006.
Yu, X., Dai, L., Demirel, S., Liu, H., & Zhang, J. (2017). Lateral load resistance of parallel bamboo strand panel-to-metal single-bolt connections–part I: yield model. Wood and Fiber Science, 49(4), 424–435.
Yu, Y., Liu, R., Huang, Y., Meng, F., & Yu, W. (2017). Preparation, physical, mechanical, and interfacial morphological properties of engineered bamboo scrimber. Construction and Building Materials, 157, 1032–1039.
Zhang, H., Pizzi, A., Zhou, X., Lu, X., & Wang, Z. (2017). The study of linear vibrational welding of moso bamboo. Journal of Adhesion Science and Technology, 1–10.
Zhang, X., Li, J., Yu, Z., Yu, Y., & Wang, H. (2017). Compressive failure mechanism and buckling analysis of the graded hierarchical bamboo structure. Journal of materials science, 52(12), 6999–7007.
Zhang, Z.W., Li, Y.S., Liu, R., & Zhang, J.L. (2017). Progressive failure of bamboo-steel adhesive bonding interface subjected low-energy impact and tension in sequence. Journal of Adhesion Science & Technology, 32(7), 1–18.
Zhong, Y., Wu, G., Ren, H., & Jiang, Z. (2017). Bending properties evaluation of newly designed reinforced bamboo scrimber composite beams. Construction and Building Materials, 143, 61–70.

Bakar, E.S., Nazip, M.N.M., Anokye, R., & Seng Hua, L. (2018). Comparison of three processing methods for laminated bamboo timber production. Journal of Forestry Research.

Chang, F.-C., Chen, K.-S., Yang, P.-Y., & Ko, C.-H. (2018). Environmental benefit of utilizing bamboo material based on life cycle assessment. Journal of Cleaner Production, 204, 60–69.

Chen, G., Luo, H., Wu, S., Guan, J., Luo, J., & Zhao, T. (2018). Flexural deformation and fracture behaviors of bamboo with gradient hierarchical fibrous structure and water content. Composites Science & Technology, 157, 126–133.

Chen, G., Luo, H., Yang, H., Zhang, T., & Li, S. (2018). Water effects on the deformation and fracture behaviors of the multi-scaled cellular fibrous bamboo. Acta Biomaterialia, 65, 203–215.

Chen, M., & Fei, B. (2018). In-situ Observation on the Morphological Behavior of Bamboo under Flexural Stress with Respect to its Fiber-foam Composite Structure. Bioresources, 13(3), 5472–5478.

Cui, Z., Xu, M., Chen, Z., & Xiang, J. (2018). Experimental study on thermal performance of bamboo scrimber at elevated temperatures. Construction & Building Materials, 182, 178–187.

Fernanda Garcia-Aladin, M., Francisco Correal, J., & Jaime Garcia, J. (2018). Theoretical and experimental analysis of two-culm bamboo beams. Proceedings Of the Institution Of Civil Engineers-Structures And Buildings, 171(4), 316–325.

Fu, G., Yitao, Z., Wujie, Z., Xing, Y., Deng, P., & Mun, W.A.S., et al. (2018). Can bamboo fibres be an alternative to flax fibres as materials for plastic reinforcement? a comparative life cycle study on polypropylene/flax/bamboo laminates. Industrial Crops and Products, 121, 372–387.

Guan, X., Yin, H., Liu, X., Wu, Q., & Gong, M. (2018). Development of lightweight overlaid laminated bamboo lumber for structural uses. Construction And Building Materials, 188, 722–728.

Hao, H., Tam, L.-h., Lu, Y., & Lau, D. (2018). An atomistic study on the mechanical behavior of bamboo cell wall constituents. Composites Part B-Engineering, 151, 222–231.

He, S., Xu, J., Wu, Z.X., Yu, H., Chen, Y.H., & Song, J.G. (2018). Effect of bamboo bundle knitting on enhancing properties of bamboo scrimber. European Journal of Wood and Wood Products, 76(3), 1071–1078.

Hector, A., Sebastian, K., David, T., Edwin, Z.E., & Harries, K.A. (2018). Bamboo reinforced concrete: a critical review. Materials and Structures, 51(4), 102–.

Hsu, C.-Y., Yang, T.-C., Wu, T.-L., Hung, K.-C., & Wu, J.-H. (2018). Effects of a layered structure on the physicomechanical properties and extended creep behavior of bamboo-polypropylene composites (BPCs) determined by the stepped isostress method. Holzforschung, 72(7), 589–597.

Hu, Y.a., He, M., Hu, X., Song, W., Chen, Z., Yu, Y., Yu, W. (2018). Bonding Technology for Bamboo-based Fiber Reinforced Composites with Phyllostachys bambusoides f.shouzhu Yi. Bioresources, 13(3), 6047–6061.

Huang, P., Chang, W.-s., Ansell, M.P., Bowen, C.R., Chew, J.Y.M., & Adamaki, V. (2018). Thermal and hygroscopic expansion characteristics of bamboo. Proceedings of the Institution of Civil Engineers-Structures And Buildings, 171(6), 463–471.

Jeong, D.S., Han, S.H., & Kim, Y.C. (2018). Effects of Heat Treatment on the Physical Properties of PP Composites with Bamboo Fiber Treated by Silane. Polymer-Korea, 42(6), 960–966.

Khoshbakht, N., Clouston, P.L., Arwade, S.R., & Schreyer, A.C. (2018). Computational modeling of laminated veneer bamboo dowel connections. Journal of materials in civil engineering, 30(2).

Kumar, G., & Ashish, D.K. (2018). Analysis of stress dispersion in bamboo reinforced wall panels under earthquake loading using finite element analysis. Computers And Concrete, 21(4), 451–461.

Li, H., Wu, G., Zhang, Q., Deeks, A.J., & Su, J. (2018). Ultimate bending capacity evaluation of laminated bamboo lumber beams. Construction And Building Materials, 160, 365–375.

Li, J., Chen, Y., Xu, J., Ren, D., Yu, H., Guo, F., & Wu, Z. (2018). The Influence of Media Treatments on Color Changes, Dimensional Stability, and Cracking Behavior of Bamboo Scrimber. International Journal of Polymer Science.

Liang, F., Wang, R., Hongzhong, X., Yang, X., Zhang, T., & Hu, W., et al. (2018). Investigating pyrolysis characteristics of moso bamboo through tg-ftir and py-gc/ms. Bioresource Technology, 256, 53–60.

Ma, X., Shi, S.Q., Wang, G., Fei, B., & Jiang, Z. (2018). Long creep-recovery behavior of bamboo-based products.

Mali, P.R., & Datta, D. (2018). Experimental evaluation of bamboo reinforced concrete slab panels. Construction and Building Materials, 188, 1092–1100.

Mannan, S., Parameswaran, V., & Basu, S. (2018). Stiffness and toughness gradation of bamboo from a damage tolerance perspective. International Journal of Solids and Structures, S0020768318301264.

Meng, J., & Sun, D.G. (2018). Effects of node distribution on bending deformation of bamboo. Journal of Tropical Forest Science, 30(4), 554–559.

Munis, R.A., Camargo, D.A., de Almeida, A.C., de Araujo, V.A., de Lima Junior, M.P., Morales, E.A.M., Cortez-Barbosa, J. (2018). Parallel Compression to Grain and Stiffness of Cross Laminated Timber Panels with Bamboo Reinforcement. Bioresources, 13(2), 3809–3816.

Niu, X., Pang, J., Cai, H., Li, S., Le, L., & Wu, J. (2018). Process Optimization of Large-Size Bamboo Bundle Laminated Veneer Lumber (BLVL) by Box-Behnken Design. Bioresources, 13(1), 1401–1412.

Okahisa, Y., Kojiro, K., Kiryu, T., Oki, T., Furuta, Y., & Hongo, C. (2018). Nanostructural changes in bamboo cell walls with aging and their possible effects on mechanical properties. Journal of Materials Science, 53(6), 3972–3980.

Penellum, M., Sharma, B., Shah, D.U., Foster, R.M., & Ramage, M.H. (2018). Relationship of structure and stiffness in laminated bamboo composites. Construction and Building Materials, 165, 241–246.

Shah, D.U., Sharma, B., & Ramage, M.H. (2018). Processing bamboo for structural composites: influence of preservative treatments on surface and interface properties. International Journal of Adhesion and Adhesives, S0143749618301246.

Sharma, B., Shah, D.U., Beaugrand, J., Janeček, E.R., Scherman, O.A., & Ramage, M.H. (2018). Chemical composition of processed bamboo for structural applications. Cellulose.

Silva Brito, F.M., Paes, J.B., da Silva Oliveira, J.T., Chaves Arantes, M.D., Vidaurre, G.B., & Brocco, V.F. (2018). Physico-mechanical characterization of heat-treated glued laminated bamboo. Construction And Building Materials, 190, 719–727.

Srivaro, S., Rattanarat, J., & Noothong, P. (2018). Comparison of the anatomical characteristics and physical and mechanical properties of oil palm and bamboo trunks. Journal of Wood Science.

Takeuchi, C.P., Estrada, M., & Linero, D.L. (2018). Experimental and numerical modeling of shear behavior of laminated Guadua bamboo for different fiber directions. Construction And Building Materials, 177, 23–32.

Teng-Chun, Y., & Tung-Ying, L. (2018). Effects of density and heat treatment on the physico-mechanical properties of unidirectional round bamboo stick boards (ubsbs) made of makino bamboo (phyllostachys makinoi). Construction and Building Materials, 187, 406–413.

Wang, C.L., Liu, Y., & Zhou, L. (2018). Experimental and numerical studies on hysteretic behavior of all-steel bamboo-shaped energy dissipaters. Engineering Structures, 165, 38–49.

Wen, K., Bu, C., Liu, S., Li, Y., & Li, L. (2018). Experimental investigation of flexure resistance performance of bio-beams reinforced with discrete randomly distributed fiber and bamboo. Construction & Building Materials, 176, 241–249.

Wu, Y., & Xiao, Y. (2018). Steel and glubam hybrid space truss. Engineering Structures, 171, 140–153.

Xu, Q., Leng, Y., Chen, X., Harries, K.A., Chen, L., & Wang, Z. (2018). Experimental study on flexural performance of glued-laminated-timber- bamboo beams. Materials And Structures, 51(1).

Yang, X., Zhang, T., Jiang, C., Wang, J., Fei, B., Liu, Z., & Jiang, Z. (2018). Directional Laminated Thermally Modified Bamboo: Physical, Mechanical, and Fire Properties. Bioresources, 13(3), 5883–5893.

Ye, F., & Fu, W. (2018). Physical and mechanical characterization of fresh bamboo for infrastructure projects. Journal of Materials in Civil Engineering, 30(2), 05017004.

Yu, H., Du, C., Huang, Q., Yao, X., Hua, Y., Zhang, W., Liu, H. (2018). Effects of Extraction Methods on Anti-Mould Property of Bamboo Strips. Bioresources, 13(2), 2658–2669.

Zhang, H., Li, H., Corbi, I., Corbi, O., Wu, G., Zhao, C., & Cao, T. (2018). AFRP Influence on Parallel Bamboo Strand Lumber Beams. Sensors, 18(9).

Zhang, J., Li, Y., Liu, R., Xu, D., & Bian, X. (2018). Examining bonding stress and slippage at steel-bamboo interface. Composite Structures, 194.

Zhao, W., Zhou, J., Long, Z., & Peng, W. (2018). Compression performance of thin-walled square steel tube/bamboo plywood composite hollow columns with binding bars. Advances In Structural Engineering, 21(3), 347–364.

Zhou, A., Bian, Y., Shen, Y., Huang, D., & Zhou, M. (2018). Inelastic Bending Performances of Laminated Bamboo Beams: Experimental Investigation and Analytical Study. Bioresources, 13(1), 131–146.

Zhou, A., Huang, Z., Shen, Y., Huang, D., & Xu, J. (2018). Experimental Investigation of Mode-I Fracture Properties of Parallel Strand Bamboo Composite. Bioresources, 13(2), 3905–3921.

Zhou, J., Huang, D., Ni, C., Shen, Y., & Zhao, L. (2018). Experiment on Behavior of a New Connector Used in Bamboo (Timber) Frame Structure under Cyclic Loading. Advances In Materials Science And Engineering.

Zujian, H., Yimin, S., & Florian, M. (2018). Hygrothermal performance of natural bamboo fiber and bamboo charcoal as local construction infills in building envelope. Construction and Building Materials, 177, 342–357.

Modern Engineered Bamboo Structures – Xiao, Li & Liu (eds)
© 2020 Taylor & Francis Group, London, ISBN 978-1-138-35185-1

State-of-the-art of practice in Colombia on engineered Guadua bamboo structures

J.F. Correal
Universidad de Los Andes, Bogotá, Colombia
Colombian Association for Earthquake Engineering (AIS), Bogotá, Colombia

ABSTRACT: *Guadua Angustifolia Kunt* (Guadua) is a bamboo species that has been considered an alternative material for construction in some of the Latin-American countries. In Colombia, specially in some Department or State like Quindio, Risaralda, Caldas, Antioquia, Tolima, Cauca, Valle del Cauca, Choco and Cundinamarca, Guadua has been used mainly for residential buildings, having bahareque (Split bamboo and mortar connected to a *guadua* frame) as the preferred structural system for this type of structures. This system has shown an adequate seismic behavior during earthquakes like the one in *Armenia* (1999), (EERI, 2000). A presidential decree in 2002, incorporated into the second version of the Colombian Building Code (NSR-98) requirements for the design and construction of two story bahareque houses. In 2010, an update of the Colombian Building Code (NSR-10) was conducted and for the first time in Colombia a new chapter about requirements for the design and construction of engineered Guadua bamboo structures was introduced. This paper presents a summary about the state of the art of practice in Colombia on engineered Guadua Bamboo structures based on the requirements stated in the Colombian Building Code (NSR-10).

1 INTRODUCTION

Bamboo is a woody perennial grass that belongs to the Poaceae family. In the world there are around 1,100 bamboo species, 451 of them are located in tropical America (Castaño et al., 2004). *Guadua* is a gigantic bamboo specie, being the biggest one in America and third biggest in the world. The stem or culm is the visible part of the *Guadua*, it has a conical cylinder form, which is divided by nodes, the distance between the nodes varies along the culm and the cavity between two nodes is known as internode. The diameter of the *Guadua* vary typically between 10 cm and 12 cm along the culm and its thicknesses also vary from 0.6 cm to 2 cm. *Guadua* has been considered an alternative earthquake-resistant material for structural purposes in countries like Colombia, Mexico, Costa Rica, Peru and Ecuador. In Colombia, especially in the coffee zone, *Guadua* has been used in one and two story dwellings. *Bahareque* is the most used structural system for this type of buildings; this is a shear wall-based system that consists in guadua frames covered with "*esterilla*" (split *guadua culm*) panels and a steel mesh with mortar topping.

Due to the good seismic performance during earthquake in *Armenia* (1999), (EERI, 2000), the Colombian Building Code Committee suggested to include requirements for the design and construction for two story bahareque houses in the current Colombian Building Code (NSR-98). Through a presidential decree in 2002, provisions to design and construction up to two story houses made of *Bahareque* were incorporated into the Colombian Building Code, NSR-98, (AIS,1998). These provisions were introduced in a unique chapter (Chapter. E) that contained simplified design and construction requirements to build until two story houses. This chapter of the NSR-98 covers three specific topics: 1) general provisions for *Bahareque*, 2) floors and joints, 3) roofs and 4) appendix about validation of structural strength of shear walls made of *Bahareque.*

During the update of the Colombian Building Code in 2010, NSR-10 (AIS, 2010), a new Chapter (Chapter G. 12) regarding design and construction of engineered *Guadua* bamboo structures was developed. This chapter covers: general requirements of *Guadua* bamboo (scope and definitions), materials, fundamental for structural design and methods, design of elements subject to bending, axial, combined forces, joints and fabrications and construction. This paper shown a summary of the main provisions stated in the NSR-10 which is considered the state of art of practice in Colombia on engineered Guadua Bamboo structures.

2 ONE AND TWO STORY HOUSES MADE OF CEMENTED BAHAREQUE

Chapter E of NSR-10 provided requirements for the design of one and two story houses made on masonry and cemented bahareque. This chapter provided simplified design requirements, thus a structural engineering is not required to perform the design.

Particularly, bahareque shear walls is a system comprised of two main parts: the framing and the sheathing (Figure 1). The framing is comprised of two horizontal elements at top and bottom, and vertical elements, connected by nails, screws or other type of fastener. The cross section of the top and bottom elements must have a minimum width equal to the diameter of the bamboo elements used as vertical, and a depth greater than 100 mm. In some cases, it is preferred to construct both top and bottom elements in sawn lumber since its connections provide greater stiffness and are less susceptible to crushing than bamboo elements. The studs may be constructed with dry bamboo with a diameter not less than 80 mm. The horizontal spacing between bamboo studs need not be less than 300 mm and should not exceed 600 mm on center.

Walls of cemented bahareque for low-rise bamboo buildings are classified into three groups. First walls are those with diagonal elements, which can be considered as structural walls comprised of top and bottom plates, diagonal bamboo elements and sheathing with split bamboo mat, covered with cement mortar applied over a steel mesh. These walls receive vertical loads and in-plane horizontal wind or seismic loads and are part of the lateral force resisting system in bamboo buildings. The second group is the load bearing walls without diagonal elements. This wall may be used only for resisting vertical loads of the building. The third group of walls are the non-bearing walls. This

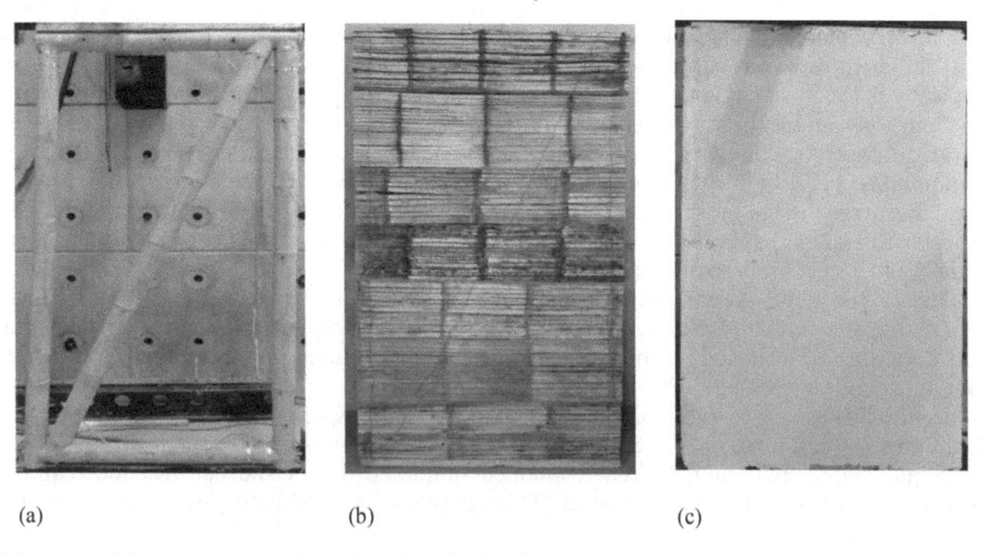

(a) (b) (c)

Figure 1. Bahareque shear wall: (a) Framing; (b) sheathing; (c) mortal covering.

Table 1. Density coefficient values for bahareque walls (adapted from AIS, 2010).

Effective peak acceleration, A_a (g)	Wall density coefficient, C_B
0.40	0.32
0.35	0.28
0.30	0.24
0.25	0.20
0.20	0.16
0.15	0.16
0.10	0.16
0.05	0.16

wall are capable only of supporting their self-weight. Exterior or interior non-bearing walls must be connected to the top diaphragm by a connection that prevents overturning, but does not allow the transmission of shear or vertical forces between the roof or floor and the wall.

Structural walls necessity be continuous from the foundation level to the top diaphragm to which they are connected. Based on NSR-10, the bahareque wall sheathing can be either split bamboo mat (or wood panel) or horizontal wood strips (Figure 1b), both covered with cement mortar applied over a steel mesh (Figure 1c). In both cases, it is recommended to use screws to attach the split bamboo mat or wood strips to the bamboo frame instead of nails. Mortar covering should be applied over a steel wire mesh having a maximum diameter of 1.25 mm (BWG 18 gauge) and hexagonal openings not larger than 25.4 mm. The minimum required classification for the cement mortar should correspond to type N (based on NSR-10), with a volume ratio not exceeding 4 parts sand to one-part cementitious material.

NSR-10 specifies a minimum length of bahareque walls in each principal direction of the building in order to provide a uniform resistance to seismic forces in the inelastic range (Equation 1). Equation 1 is developed for bahareque walls sheathed on both sides. The effective length of a wall with one side sheathed is taken to be one half of its actual length.

$$L_i \geq C_B A_P \tag{1}$$

Where L_i is the minimum total length of walls without openings in each direction, C_B is a wall density coefficient determined from Table 1 as a function of the effective peak acceleration A_a (based on seismic hazard maps), and A_p is the area tributary to the story for which the walls are designed. For second story walls in two-story houses, $0.66A_p$ may be used if light-weight materials are used for the roof, such as panels or metallic sheets without mortar covering. Bahareque walls must be distributed in an approximately symmetric manner in plan.

Cemented bahareque wall systems must be designed for the effects of combined loading. For the effects caused by seismic forces, the energy dissipation capacity of the structural system is taken into account, applying the seismic response modification factor (R). NSR-10 specifies an R-factor of 1.5 for cemented bahareque. Values obtained for stresses over each wall, due to vertical and lateral loads, must be less than the allowable strength. Table 17.8, adapted from NSR-10, establishes values for allowable design forces and stresses according to the structural composition and materials used for each type of bahareque wall. The values expressed in Table 17.8 assume that walls will be fully anchored to the foundation and between each other.

Table 2 . Allowable strength values for Guadua a.K. (adapted from AIS, 2010).

Property	Notation	Allowable strength (MPa)
Bending	F_b	15
Tension	F_t	18
Compression-parallel-to-grain	F_c	14
Compression-perpendicular-to-grain	F_p	1.4
Shear	F_v	1.2
Modulus of elasticity (mean)	$E_{0.5}$	9500
Modulus of elasticity (5th percentile)	$E_{0.05}$	7500
Modulus of elasticity (minimum)	E_{min}	4000

3 DESIGN AND CONSTRUCTION OF ENGINEERED GUADUA BAMBOO STRUCTURES

The Chapter G. 12 of the NSR-98 established requirements for the design and constructions of engineered Guadua bamboo Structures. A summary of the main requirements are presented as follows.

Additional information can be find in NSR-10 and in Chapter 14 of the Nonconventional and Vernacular Construction Materials (Correal, 2016).

3.1 *Fundamental for structural design*

NSR-10 established that bamboo elements are considered homogeneous and linear for the estimation of stresses due to applied loads. Also, the analysis and design of bamboo structures must be based on the principles of engineering mechanics and strength of materials. In addition, the analysis must reflect all the possible acting loads over the structure during the construction and service stages.

Regarding the design of earthquake resisting buildings made of Guadua, the seismic response coefficient (R) is needed to reduce the linear elastic design seismic forces to account for the energy dissipation capacity of the structure. The NSR-10 recommends a basic seismic design coefficient (R_0) for Guadua structures, whose lateral force resisting system is based on diagonally braced walls, to be equal to 2.0. For the case of a lateral force resisting system based on bahareque walls, the value of R_0 is 1.5.

The ASD design philosophy is used by NSR-10. Thus, allowable stress is determined by factoring the characteristic value of the Guadua by a series of modification factors (which include such effects as the test load rate, load duration and safety factor). Once the characteristic value is determined for each type of load action (bending, tension, compression and shear), allowable stresses is determined. Allowable stresses used in NSR-10 for Guadua are shown in Table 2.

Stress values used in the design of different elements of structures, such as beams and columns, are adjusted according to the following equation (NSR-10 Equation G. 12.7-3):

$$F_i' = F_i C_D C_m C_t C_L C_r C_c \tag{2}$$

Where the adjustment factors are for load duration (C_D), moisture content (C_m), temperature (C_t), beam lateral stability (C_L), group action (C_r), and shear (Cc). Chapter G. 12.7 of the NSR-10 presents the values for adjustment factors for Guadua bamboo, which are reproduced in Table 3.

Table 3. Adjustment factors for Guadua a.K. (adapted from AIS, 2010).

			F_b	F_t	F_c	F_p	F_v	$E_{0.5}$	$E_{0.05}$	E_{min}
C_D	Permanent		0.90	0.90	0.90	0.90	0.90	-	-	-
	Ten years		1.00	1.00	1.00	0.90	1.00	-	-	-
	Two months		1.15	1.15	1.15	0.90	1.15	-	-	-
	Seven days		1.25	1.25	1.25	0.90	1.25	-	-	-
	Ten minutes		1.60	1.60	1.60	0.90	1.60	-	-	-
	Impact		2.00	2.00	2.00	0.90	2.00	-	-	-
C_m	MC ≤ 12%		1.00	1.00	1.00	1.00	1.00	1.00	1.00	1.00
	MC = 13%		0.96	0.97	0.96	0.97	0.97	0.99	0.99	0.99
	MC = 14%		0.91	0.94	0.91	0.94	0.94	0.97	0.97	0.97
	MC = 15%		0.87	0.91	0.87	0.91	0.91	0.96	0.96	0.96
	MC = 16%		0.83	0.89	0.83	0.89	0.89	0.94	0.94	0.94
	MC = 17%		0.79	0.86	0.79	0.86	0.86	0.93	0.93	0.93
	MC = 18%		0.74	0.83	0.74	0.83	0.83	0.91	0.91	0.91
	MC ≥ 19%		0.70	0.80	0.70	0.80	0.80	0.90	0.90	0.90
C_t	T ≤ 37°C		1.00	1.00	1.00	1.00	1.00	1.00	1.00	1.00
	37 °C < T ≤ 52°C	MC ≥ 19%	0.60	0.85	0.65	0.80	0.65	0.80	0.80	0.80
		MC < 19%	0.85	0.90	0.80	0.90	0.80	0.90	0.90	0.90
	52 °C < T ≤ 65°C	MC ≥ 19%	0.40	0.80	0.40	0.50	0.40	0.80	0.80	0.80
		MC < 19%	0.60	0.80	0.60	0.70	0.60	0.80	0.80	0.80
C_L	D/b = 2		1.00	-	-	-	-	-	-	-
	D/b = 3		0.95	-	-	-	-	-	-	-
	D/b = 4		0.91	-	-	-	-	-	-	-
	D/b = 5		0.87	-	-	-	-	-	-	-
C_r	s ≤ 0.60 m		1.10	1.10	1.10	1.10	1.10	-	-	-
Cc	1/D = 5		-	-	-	-	-	0.70	-	-
	1/D = 7		-	-	-	-	-	0.75	-	-
	1/D = 9		-	-	-	-	-	0.81	-	-
	1/D = 11		-	-	-	-	-	0.86	-	-
	1/D = 13		-	-	-	-	-	0.91	-	-
	1/D = 15		-	-	-	-	-	0.93	-	-

4 ELEMENTS DESIGN

4.1 *Elements subjected to bending*

The design moment in bamboo is obtained using traditional elasticity theory and the corresponding stresses are calculated using the elastic section modulus, S (Equation 3). The design bending stresses, f_b, should not be higher than the maximum allowable stress for bending, F'_b, as shown in Equation 3, which are modified by appropriate adjustment factors.

$$f_b = \frac{M}{S} \leq F'_b \tag{3}$$

Where M is the flexural moment acting over the element and S is the elastic section modulus.

Lateral stability factors need to be considered in bending design of beams. Framing conditions could provide an effective connection of roof or floor diaphragms to the compression side of beam causes the unbraced length to be essentially zero, thus lateral instability is mitigated. In the case of slender bamboo beams (beams comprised of two or more bamboo culms), the lateral support of the compression side must be verified. Two methods could be used to determine the stability factor (CL): a rule-of-thumb method, or tabulated lateral stability factors prescribed by NSR-10.

For the rule-of-thumb method, the lateral stability factor CL is 1.0, provided that the following conditions are satisfied for each case of beams comprised of two or more Guadua a. K. bamboo culms:

- For D/b = 2, no lateral support is required;
- For D/b = 3, lateral displacement at the supports must be restrained;
- For D/b = 4, lateral displacement at the supports and the compression side of the beam must be restrained;
- For D/b = 5, lateral displacement at the supports must be restrained and continuous support for the compression side of the beam should be provided.

Alternatively, the tabulated lateral stability factor (CL) used in NSR-10 for built-up beams of two or more Guadua a.K. bamboo culms are presented in Table 3. These factors are related to the depth (D) to width (b) ratio of the beam.

4.2 *Elements subjected to shear*

Shear is calculated at a distance equal to the depth of the element measured from the face of the support. For beams comprised of only one culm, this depth could be equal to the outside diameter of the culm, except for cantilever beams where the maximum shear stress should be calculated at the face of the support. The maximum shear stress should be determined and must be lower than the maximum allowable shear-parallel-to-grain strength modified (Table 2) by appropriate adjustment factors (Table 3).

4.3 *Elements subjected to axial*

An axially loaded bamboo elements has the force applied parallel to the longitudinal. These forces are either tension or compression. In bamboo structures, compression force members (like columns) are encountered more often than tension members due to the natural need of the structures to carry vertical gravity loads and because of the difficulty of fabricating efficient connections for tension members.

Therefore, the most common axially-loaded members in bamboo structures are columns. There are different type of columns such as single culm columns, spaced columns and built-up columns. The main mechanisms of failure of compression elements are buckling and crushing. The slenderness ratio, λ, defines the primary measure of buckling:

$$\lambda = \frac{kl_u}{r} \tag{4}$$

Where l_u is the unbraced length of the column, and k is the effective length coefficient dependent on the end support conditions of the column, and whether side sway of the column is prevented or not and r is the radius of gyration.

According to their slenderness ratio, Guadua bamboo columns are classified into short ($kl_u/r \leq 30$), intermediate ($30 < kl_u/r < C_k$) or long columns ($C_k \leq kl_u/r \leq 150$). In which the slenderness limit between intermediate and long columns, Ck, is given by the following expression:

$$C_k = 2.56\sqrt{\frac{E_{0.05}}{F'_c}} \tag{5}$$

Where $E_{0.05}$ is the 5th percentile modulus of elasticity, and F'_c is the allowable compression-parallel-to-grain strength modified by the appropriate adjustment factors. Guadua bamboo columns with slenderness ratios greater than 150 are not allowed.

The NSR-10 defines that the acting compression stress, f_c, must not exceed the allowable compression-parallel-to-grain strength modified by the appropriate coefficients, F'_c, for short, intermediate and long columns, according to Equations 6, 7 and 8, respectively.

$$f_c = \frac{N}{A_n} \leq F'_c \tag{6}$$

$$f_c = \frac{N}{A_n\left[1 - \frac{2}{5}\left(\frac{\lambda}{C_k}\right)^3\right]} \leq F'_c \tag{7}$$

$$f_c = 3.3\frac{E_{0.05}}{\lambda^2} \leq F'_c \tag{8}$$

Where N is the load acting over the element in the direction parallel to the fibres.

Guadua bamboo elements subjected to the axial tension stress, f_t, axial must limited to the allowable strength modified by the appropriate adjustment factors, F'_c. The equation for designing bamboo tension members is:

$$f_t = \frac{T}{A_n} \leq F'_t \tag{9}$$

Where T is the tension load acting over the element having a net cross sectional area A_n.

4.4 *Elements subjected to combined loadings*

Combined loading for bamboo elements consists of bending moment that occurs simultaneously with a axial force. The combined stresses are analyzed using the Equation 10 which is applied to the stress condition at both extreme tension and compression faces of the section. For the case of net tensile stresses, the allowable bending stress used in the interaction equation, F'_b, does not include the lateral stability factor (CL = 1.0). On the other hand, if the combined stresses result in net compression, the lateral stability factor must be included in the calculation of F'_b.

$$\frac{f_t}{F'_t} + \frac{k_m f_b}{F'_b} \leq 1.0 \tag{10}$$

Where f_t and f_b correspond to the acting tension and bending stresses over the element, respectively and the amplification factor, km given by Equation 11. For combined bending and tension, $k_m = 1.0$.

$$k_m = \frac{1}{1 - 1.5(N_a/N_{cr})} \tag{11}$$

Where N_a is the acting compression force, and N_{cr} is Euler critical buckling load calculated using the 5th percentile modulus of elasticity, $E_{0.05}$.

4.5 *Joints design*

Dowel joints are one of the most widely used type of connections in bamboo construction. Bamboo dowel joints use steel bolts as a connector and usually the bamboo internodes are filled. Cement mortar is the most common filling material in the joints of bamboo buildings, since cement mortar is easier to obtain and fabricate. NSR-10 recommends using a cement mortar with a minimum cement-to-sand ratio of 1: 3 and to use plasticizer additive to ensure the fluidity of the mix. It is also recommended that the cement mortar should have a compression strength between 12.5 MPa to 17.5 MPa and 110% to 125% of fluidity.

The NSR-10 for Guadua a.K. bamboo dowel type joints recommend a minimum bolt diameter of 9.5 mm and bolts should be made of structural steel with yield strength greater than 240 MPa. Drilled holes for bolts should have a diameter 1.5 mm larger than the bolt diameter. Drilled holes for filling the internodes should have a maximum diameter of 26 mm and must be filled with cement mortar in order to ensure the structural continuity of the element. In addition, all other metallic elements used in dowel-type joints that are exposed to unfavorable environmental conditions should be treated against corrosion.

The use of metallic straps or clamps as shown in Figure 2. is permitted for the design of joints, provided appropriate precautions are taken to prevent crushing and compression-perpendicular-to-grain failure in individual culms, as well as slip between connected elements.

In the case of a dowel type joints subjected to bearing or crushing loads, the bamboo internodes adjacent to the joints, or where bolts are located, must be filled with cement mortar. Metallic washers must be used between both the bolt head and nut and the bamboo element. Dowel-type joint capacity depends on the direction of the load at the main and side elements connected. For instance, NSR-10 defines three allowable loads for dowel type joints: a) when loading in the joint is parallel to the fibres, for the main element and the side elements (indicated as P in Figure 2a); b) when the force is parallel to the fibres of the main element, but perpendicular to the fibres of the side elements, or vice versa (indicated as Q in Figure 2b); and c) when the force is perpendicular to the fibres of one element and parallel to the fibres of the other element (indicated as T in Figure 2c).

NSR-10 allowable loads for dowel-type joints subjected to double shear (P, Q and T directions) are shown in Table 4, as a function of the outside diameter (D) and the bolt diameter (d). Allowable loads listed in Table 4 are representative of Guadua a.K. bamboo culms with moisture content under 19%, used in a dry service environment. For joints of 4 or more members, each load plane should be evaluated as a single shear joint. The value for the connection should be estimated as the lowest nominal value obtained, multiplied by the number of shear planes.

For forces in the joint forms an angle with the fibre direction of the side elements, or vice versa, the allowable load should be determined from the Hankinson equation.

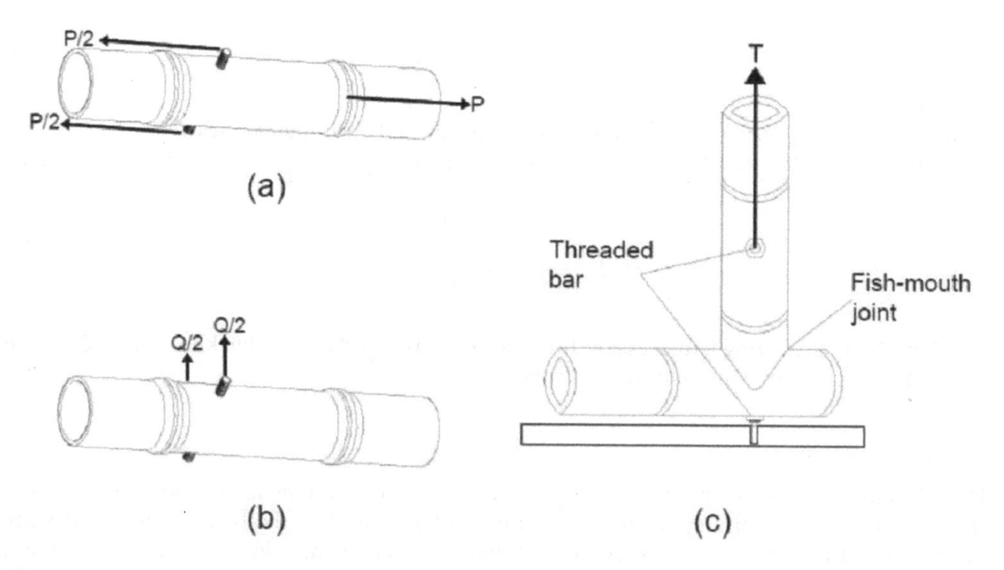

Figure 2. Guadua bamboo dowel-type connection.

Table 4. Allowable loads for dowel-type joints subjected to double shear (adapted from AIS, 2010).

Bolt diameter, d (mm)	Guadua element diameter, D_e (mm)	P (N)	Q (N)	T (N)
9.5	80	7212	2885	2000
	90	8008	3203	2100
	100	8804	3522	2200
	110	9601	3840	2300
	115	10041	4016	2400
	120	10481	4193	2500
	125	10922	4369	2600
	130	11362	4545	2700
	135	11802	4721	2800
	140	11242	4897	2900
	150	13039	5216	3000
12.7	80	9710	3884	2000
	90	9916	3966	2100
	100	10943	4377	2200
	110	11970	4788	2300
	115	12521	5009	2400
	120	13072	5229	2500
	125	13623	5449	2600
	130	14174	5670	2700
	135	14725	5890	2800
	140	15276	6110	2900
	150	16303	6521	3000
15.9	80	11540	4616	2000
	90	12806	5122	2100
	100	13250	5300	2200
	110	14515	5806	2300
	115	15185	6074	2400
	120	15855	6342	2500
	125	16525	6610	2600
	130	17195	6878	2700
	135	17865	7146	2800
	140	18535	7414	2900
	150	19800	7920	3000

Table 5. Group reduction coefficient values for connections with two or more bolts (adapted from AIS, 2010).

	Number of bolts				
Type of connection	2	3	4	5	6
Connections with Guadua elements	1.0	0.97	0.93	0.89	0.82
Connections with steel elements	1.0	0.98	0.95	0.92	0.90

Allowable loads listed in Table 4 correspond to connections with only one bolt. The group reduction coefficient, C_g, (Table 5) should only be applied to P load when a connection requires more than two bolts in a line parallel to the loading direction. Lastly, NSR-10 specifies that the spacing between bolts should not be less than 150 mm or greater than 250 mm, but in all cases an internode must fall between each bolt. The distance from a bolt to the free end of the element must be greater than 150 mm in connections subjected to tension, and 100 mm in connections subjected to compression.

5 REMARKS AND FUTURE RESEARCH

Bamboo is gaining an impressive visibility as an alternative construction material for structures, since bamboo is a renewable material with fast growing rate, high strength-to-weight ratio, low harvesting and processing cost, and carbon sequestering capability.

The Colombian Building Code (NSR-10) is without doubts one of the most advance code regarding bamboo design requirements. A summary of the main provisions stated in the NSR-10 which is considered the state of art of practice in Colombia on engineered Guadua Bamboo structures was presented. Those provisions allow to design and construct structure made of Guadua Bamboo in both non-seismic and seismic regions.

However, future research around bamboo is should focused to topics that are still uncertain or may be explored in depth, such as the determination of load and resistance factors for LRFD, the more accurate estimation of long-term deflections, the effective moment of inertia for built-up beams and columns, the structural behaviour of walls subjected to lateral loads considering various sheathing and finishing materials, or the characterization and design procedures of adequate mechanical connections (for example considering the application of widely used theoretical approaches such as the European Yield Model (EYM) to mortar-or wood-filled bamboo culms).

REFERENCES

EERI Earthquake Engineering Research Institute. 2000. El Quindío, Colombia, South America earthquake, January 25, 1999.

Castaño F. and Moreno R.D. 2004. Guadua para todos cultivo y aprovechamiento.

AIS Asociación Colombiana de Ingeniería Sísmica. 1998. Reglamento Colombiano de Construcción Sismo-Resistente NSR-98.

AIS Asociación Colombiana de Ingeniería Sísmica. 2010. Reglamento Colombiano de Construcción Sismo-Resistente NSR-10.

Correal J. 2016. Chapter 14. Bamboo design and construction. Nonconventional and Vernacular Construction Material. Edited by K.A. Harries and B. Sharma.

Modern Engineered Bamboo Structures – Xiao, Li & Liu (eds)
© 2020 Taylor & Francis Group, London, ISBN 978-1-138-35185-1

An overview of global modern bamboo construction industry: A summary report of ICBS2018

K.W. Liu
International Bamboo and Rattan Organization, Beijing, China

J. Yang
Tsinghua University, Beijing, China

R. Kaam
International Bamboo and Rattan Organization, Beijing, China

C.Z. Shao
The Chinese University of Hong Kong, HK, China
Tsinghua University, Beijing, China

ABSTRACT: Bamboo is one of the fastest growing grass species on earth and woody bamboos have been used in buildings for thousands of years. With the development of modern construction technologies, bamboo is not only used as decorative or structural material in modern buildings, but also plays important roles in transportation facilities, outdoor landscape and eco-tourism practices in rural area. Through systematic analysis and summarizing more than 50 presentations from around 30 different countries, which were delivered during Sustainable Bamboo Building Materials and Third International Conference on Modern Bamboo Structure (ICBS2018), while combing with practical experiences of projects implemented across the world by and with the support of the International Bamboo and Rattan Organization (INBAR) as well as the information from INBAR's partners, the authors provide on the one hand the state quo and trends of global modern bamboo construction industry and on the other hand presents a number of challenges and opportunities to guide the bamboo construction industry further development.

1 BACKGROUND

1.1 *BARC2018 and ICBS2018*

In June 2018, the International Bamboo and Rattan Organisation (INBAR) and the State Forestry and Grassland Administration of China co-organized the Global Bamboo and Rattan Conference (BARC2018) (Figure 1A) in Beijing. The theme of this conference was "Enhancing South-South Cooperation for Green Development through Bamboo and Rattan's Contributions to the United Nations Sustainable Development Goals (UN SDG)". At the congress, high-level dialogues, parallel meetings, symposiums and exhibitions were delivered, and attracted more than 1,200 delegates from more than 70 different countries. Prior to BARC2018, we had two successful conferences in Changsha, China in 2007, and in Bogota, Colombia in 2009, the Third International Conference on Modern Bamboo Structure (ICBS2018) provided a platform for all the stakeholders of the bamboo construction value chain to showcase the significant progress of global modern bamboo construction in the past 9 years. Along the same line, INBAR demonstrated four prefabricated bamboo houses (Figures 1B, 1C, 1D, 1F) and one prefabricated bamboo bridge (Figure 1E) on site which opened to all the public.

Figure. 1. Demonstration bamboo structures during BARC2018*.

*Bamboo and Rattan Congress 2018 in China National Convention Centre (A); Three round pole bamboo houses (B, C and D) and one bridge (E) were built by Tsinghua University; one engineered bamboo house was built by Yong'an Bamboo Industry Research Institute (F).

1.2 *Information source*

More than 200 participants from around different 30 countries attended ICBS2018, representing different stakeholder groups including officials of international organizations, researchers from universities and research institutes, architects, structural engineers, landscape designers, manufacturers, artisans, students and etc. The total number of researchers, manufacturers, and engineering personal (architects, structural engineers and landscape designers) accounted for 29%, 14% and 8% respectively, while students from 5 different countries (China, the USA, Ethiopia, Italy and Ghana) accounted for 46%. More than 50 presentations from about 40 different international organizations, universities, institutes or private sectors showcased the great progress of global modern bamboo industry which was made in different countries: the representative from the United Nations High Commissioner for Refugees (UNHCR) depicted the role of temporary shelters from the perspective of emergency relief, while the official of INBAR shared its own experience on how to promote global modern bamboo construction industry by INBAR Construction Task Force (INBAR TFC[1]). INBAR TFC is a new

1. https://www.inbar.int/focusareas/design-construction/construction-taskforce/

platform for global bamboo construction experts; Researchers from Brazil, China, Ethiopia, Ghana, Italy, the United States of America (the USA) and the United Kingdom (the UK) shared their recent research outputs; CEOs and managers from more than 20 private sectors in China, Italy, the USA, Thailand, the Netherlands, Australia, the UK, India, Nepal, Mexico and the Philippines shared those vivid bamboo construction practices in the market. The valuable information reflected the status quo and trends of global bamboo construction industry which will be given in the next chapter.

2 STATUS QUO AND TRENDS OF GLOBAL BAMBOO CONSTRUCTION INDUSTRY

2.1 *How many countries use bamboo in construction?*

According to the statistics of INBAR, more than 50 different countries use bamboo as a construction material at present (Table 1). However, the usage of bamboo is different from countries.

According to Jiang (2007), the largest distribution of bamboo forest is found in the Asian-Pacific region, followed by the Americas and Africa. In Asia, with the exception of China, Korea, Iran, Israel and Japan, all the other Asian countries mainly use round bamboo structures. Modern architecture theories and construction technologies have been applied to traditional bamboo architecture in Thailand, Indonesia and Vietnam, where lots of large-scale modern round bamboo structures were built. China is the largest exporter of engineered bamboo products in the world (INBAR, 2018). Engineered bamboo materials are mainly used as indoor and outdoor decorative materials in large and medium-size public buildings, as well as decking materials in outdoor landscape. However, engineered bamboo materials used as structural components are basically in the demonstration stage, such as small transportation facilities or one to three-storey houses.

In Latin America, Colombia has advanced construction technologies and national standardization system of round bamboo structures. Many other countries in Latin America begun to learn from Colombia in the design and construction of modern bamboo buildings, most of which are round pole bamboo structures. Engineered bamboo panels have been applied to enclosure components, such as wallboards, floors and roof panels.

Table 1. The statistics for countries who use bamboo as a construction material.

No.	Regions	Countries & Applications	
	In total: 52 countries	Round Pole bamboo Materials or Structures	Engineered bamboo Materials or Structures
1	Asia	China, Vietnam, Philippines, Indonesia, Thailand, India, Nepal, Malaysia, Cambodia, Bangladesh, Japan, Laos, Myanmar	China, South Korea, Iran, Israel, Japan
2	Latin America	Colombia, Ecuador, Peru, Mexico, Costa Rica, Brazil, Haiti, Guatemala, Argentina	Mexico, Chile, Colombia, Ecuador, Argentina
3	Africa	Madagascar, Ethiopia, Congo, Rwanda, Mozambique, Nigeria, Senegal, Ghana, Tanzania, Uganda, Kenya	South Africa
4	Europe	Netherlands, Spain, France, Germany	Netherlands, Italy, Spain, France, Germany, Belgium, Switzerland, Norway, UK, Portugal, Swiss
5	North America	/	USA
6	Oceania	/	Australia

Source: authors own tabulation based on review of bamboo construction projects in different countries.

In Africa, bamboo is mainly used for traditional structures. In a few countries, architects or engineers have started to improve traditional bamboo structures.

In European countries and Australia, they mainly use engineered bamboo materials as indoor and outdoor decorative materials. In some countries, such as Netherlands, Spain, France and Germany, round bamboo poles are used for making walls or windows in low cost social houses or parking buildings.

In the USA, engineered bamboo materials are used as indoor and outdoor decorative materials, as well as structural components.

2.2 *What are the main applications of bamboo as a construction material?*

Based on INBAR's experience on bamboo construction and combined with the information provided by all the participants during BARC2018 and other partners, bamboo is mainly used in the following aspects as a construction material at present: 1) Earthquake or natural disasters resistant houses; 2) Low cost social houses; 3) Round pole bamboo structures and engineered bamboo structures; 4) Small transportation facilities; 5) Outdoor landscape; 6) Eco-tourism practices in rural area; 7) High-end custom-made bamboo products.

2.2.1 *Earthquake or natural disasters resistant houses*

Bamboo is light-weight, high strength with excellent earthquake resistant performance. Several large earthquakes in many countries have verified this advantage of bamboo buildings. For example, the 1991 Richter 7.6 earthquake in Costa Rica caused many concrete buildings to collapse, however, 20 bamboo buildings in the epicenter of the earthquake were intact (Mary Roah, 1996). When the magnitude 6.4 earthquake struck Colombia's coffee region in 1999, it was found that many traditional Bahareque bamboo houses in the area survived, while more modern masonry and reinforced concrete buildings suffered significant damages and often collapsed (S. Kaminski et al, 2016). In 2016, following Ecuador's magnitude 7.8 earthquake, INBAR organized experts to conduct post-earthquake assessment on local buildings, and found that the earthquake had little impact on local bamboo structures. In 2018, an earthquake of magnitude 6.9 occurred in the Prawira area of Indonesia. Most of the concrete buildings such as mosques, were close to collapse (VOA News, 2018). However, traditional houses built with natural materials such as bamboo, wood and reed still stood firmly. The above examples show that the bamboo house has excellent seismic performance which should be promoted in earthquake prone regions.

Many countries also use bamboo as an emergency relief material after earthquake. After the 2016 earthquake in Nepal, INBAR cooperated with a local architectural design company "ABARI" to develop and build 3 brick-concrete school houses with bamboo roofs (Figure 2A, 2B), 6 bamboo-earth dwellings (Figure 2C), and 10 traditional bamboo houses for school rooms (Figure 2D). Ms.Verónica María Correa Giraldo[2], an expert of INBAR TFC, participated in an emergency relief project after the Mexican earthquake in 2017, which provided 17 temporary bamboo shelters (Figures 3A, 3B, 3C, 3D) for local residents after the earthquake. So far people still live in those temporary bamboo shelters.

2.2.2 *Low cost social houses*

According to the report "World Urbanization Prospects: The 2018 Revision" published by the United Nations, 55% of the world's population currently lives in urban areas, this is expected to increase to 68% by 2050. In the process of rapid urbanization, one out of eight people lives in slums, and the housing conditions of the poor need to be improved urgently. However, bamboo has excellent performance and is low price. It can partly replace other industrial building materials (steel, concrete and cement) and natural materials (wood and stone) in countries and regions which are rich in bamboo resources to provide basic and decent housing for the poor and disadvantaged people. In Ecuador, the Philippines and Indonesia, the development and application of bamboo social housing has greatly alleviated the pressure of urbanization. In Ecuador, a non-

2. https://www.inbar.int/tf_veronica-giraldo/

Figure 2. Reconstruction project after earthquake in Nepal (all photos are provided by INBAR):Brick-concrete houses with bamboo roofs (A and B); Bamboo-earth house (C); Traditional bamboo house (D).

Figure 3. Temporary bamboo shelters for emergency relief (Photos are provided by Verónica María Correa Giraldo from Kaltia & Bambuterra).

governmental organization called Viviendas del Hogar de Cristo (VHC) has been using prefabricated bamboo boards and wooden frames to build social housing for the poor since the 1970s. Over the past 47 years, a total of 200,000 families have been provided with decent housing, mainly benefiting social groups living in slums or street evictions. These affordable housing, which covers an area of more than 30 square meters and costs about $1,000 per set of bamboo houses, provide a reliable shelter for the poor (VHC, 2018) (Figure 4A). Philippines is trying to use earthquake-resistant Bahareque bamboo housing technology from South America and the modern engineering technology from Europe with the traditional skills of the Philippines. Over the past two years, a local company called BASE Bahay has provided nearly 400 bamboo houses (Figure 4B, 4C) to more than 3,000 local poor people (BASE, 2018). The average cost per set of bamboo house is around $140 per square meters[3]. These bamboo houses not only have excellent seismic performance, but could also effectively resist frequent typhoons. In Indonesia, Danone, an international food product company, commissioned the Pemulung House to create healthy, well organized housing compounds for garbage collectors in Bali. The bamboo houses (Figure 4D) are created as modules with main living spaces on the first floor and a mezzanine sleeping area above. Room for safe storage of recycled materials was also integrated into the design (IBUKU, 2011).

2.2.3 *Modern round pole bamboo structures or engineered bamboo structures*

Owing to its unique natural shape and excellent structural performance, round pole bamboo has been used as a structural component for thousands of years. Since the 1970s, architects and engineers started to use modern technologies and architectural theories to construct modern bamboo buildings (Liu et al, 2013). With over 40 years of development, modern round pole bamboo structures have been successfully used in modern buildings, such as cathedral, museum, hotel and school building, with large spans and voluminous spaces in many tropical countries like Colombia, Thailand, Indonesia and Vietnam. For example, the Bamboo Sports Hall of Panyaden International School (Figures 5A, 5B) designed by Chiangmai Life Architects and Construction[4] and completed in 2017 in Thailand. The hall covers an area of 782 square meters with a capacity of 300 students. The structure utilizes prefabricated

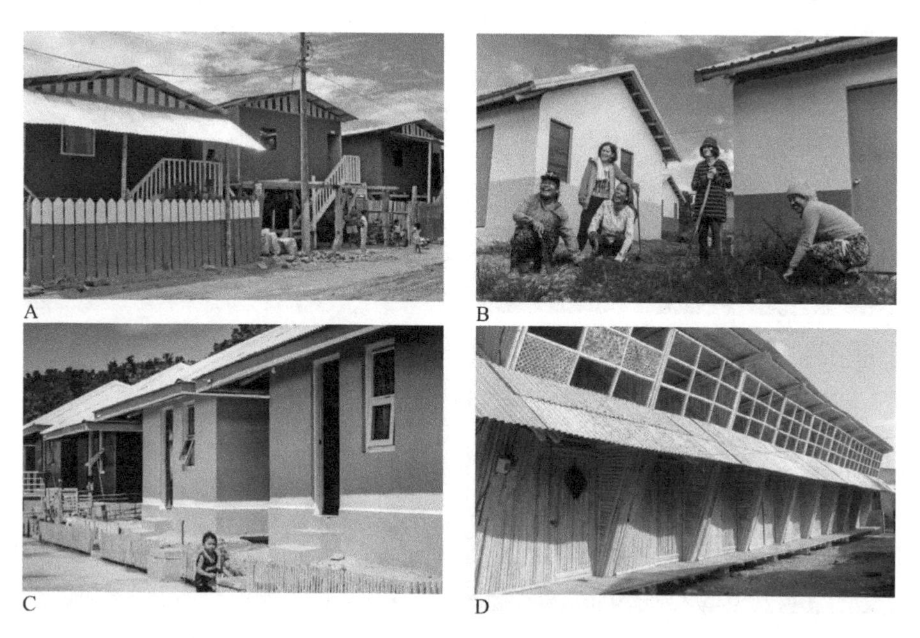

Figure 4. Social bamboo houses (A, B and C are bamboo houses in Ecuador, Philippines, Indonesia and Mexico).

Photos source: A (https://hogardecristo.org.ec (retrieve on 30th November, 2018)); B and C provided by Luis Felipe from BASE Bahay Company; D (http://ibuku.com/pemulung-housing/(retrieve on 30th November, 2018)).

---

3. Information provided by Luis Felipe from Base Bahay Foundation.
4. https://www.bamboo-earth-architecture-construction.com; the detailed information of Bamboo Sports Hall is provided by Markus Roselieb from Chaingmai Architects and Construction

Figure 5. Round bamboo structures and engineered bamboo structures.

Photos source: Bamboo Sports Hall in Thailand (A and B, provided by Markus Roselieb from Chiangmai Life Architects and Construction); Contemplation Bamboo Pavilion in Arles, France (C and D, provided by Photographer Xavier de Jauréguiberry); Three-storey laminated bamboo structure in Nanjing, China (E, provided by Dongsheng Huang from Nanjing Forestry University); Public toilet in Nanjing, China (F, provided by Eric Xiong from Greenzu Bamboo Company); A high-end residence roof in South Carolina, the USA (G, provided by Jeran Hammann from Lamboo); A hybrid steel and bamboo ceiling system of the airport terminal in Illinois, the USA (H, provided by Jeran Hammann from Lamboo).

bamboo trusses to span more than 17 meters without steel reinforcements or connections (Designboom, 2018). The design does not only enable a cool and pleasant climate all year through natural ventilation and insulation, but also meets the requirements of modern safety standards to withstand various loads, including local high-speed winds, earthquakes and all other natural forces. Worth noting, modern round pole bamboo structures are not only used in tropical areas, but also boomed in countries where bamboo does not grow naturally. For example, a Contemplation Bamboo Pavilion (Figures 5C, 5D) was built in 2018 in Arles, France. The local engineering and architecture company C&E cooperated with Simón Vélez, one of the most famous architects in bamboo construction in the world, to design a bamboo structure for photography exhibition. The pavilion is 68 meters long, 16 meters wide and 9 meters high and is made up of a set of Guadua bamboo frame arranged according to a 2.5 center distance assembled by steel framework hoops at variable heights to make it extremely rigid (C&E, 2018).

Engineered bamboo is a very new market relative to round pole bamboo. Engineered bamboo materials, such as laminated bamboo or bamboo scrimber, have properties which could compare with or surpass those of timber (Bhavna, 2015) and have great potential to be used as structural components in buildings. However, due to the lack of national or international standards, coupled to the fact that different countries have different policies for new structural materials, the structural uses of engineered bamboo materials have not been scaled up in the market. For example, in China, engineered bamboo structures are almost built on non-commercial lands with the purpose of demonstration or to be used as facilities in scenic area. Figure 5E shows a three-storey building constructed by laminated bamboo in 2017 which is located in the Teaching and Scientific Research Base of Nanjing Forestry University in Nanjing, China. The average price of the main structure is around $300 per square meter, for which only four workers took one month to finish the construction. Another 300 square meters public toilet (Figure 5F) located in a park of Nanjing, China was also built by laminated bamboo mainly used as columns, beams, rafters, wallboards and grilles in the building. All the structural bamboo components were prefabricated in the factories and installed on site that greatly shorted the construction period as well as reduced the construction cost.

The situation in the USA is much better than other countries. Since American Society for Testing Materials (ASTM) announced an imminent revision to D5456: Standard Specification for Evaluation of Structural Composite Lumber Products that "would include bamboo as a fiber material that can be used in the manufacture of products covered in the standard." in 2010 (C.C. Sullivan, 2018), ASTM D5456 becomes the first standard to recognize laminated veneer bamboo as a structural product and provides guidance on manufacturing standards and test methods. The bamboo material is treated as an equivalent to structural composite lumber products such as laminated strand lumber, laminated veneer lumber, oriented strand lumber and parallel strand lumber. (Gatóo et al, 2014) For example, solid curved engineered bamboo beams were used as the main roof support for a high-end residence located in South Carolina, USA. These beams measure at 22 centimeters wide, 42 centimeters deep, 11.3 meters long with a 8.2 meters camber (Figure 5G). Another project utilized a ceiling system to compliment the curved beams. Engineered bamboo beams used as cross bracing between the structural iron in an airport terminal in Marion, Illinois, USA (Figure 5H). Currently, some of the main challenges that bamboo promoters in the USA are facing is lack of knowledge and capacity to designs structures and demonstrated to the public that these bamboo products actually exist and can easily be integrated into most areas where other traditional natural materials such as wood and steels are used.[5]

2.2.4 *Small transportation facilities*

With the exception of traditional round pole bamboo bridges, both round bamboo poles and engineered bamboo materials are applied to build small transportation facilities in recent years. For example, a lightweight and inexpensive toll station was built

5. The detailed information of two practices and the experience on the structural use of bamboo in the USA are provided by Jeran Hammann from Lamboo

Figure 6. Bamboo used in transportation facilities.

Photos source: Toll station in Colombia (A, photographer Kewei Liu from INBAR); Parking slot in China (B, provided by Weiren Zeng from Anji Bamboo Research and Design Center, China); Parking slot in South Africa (C, taken by BMW Group South-Africa and provided by José Luken from Moso International); Traffic sign in Netherlands (D, provided by José Luken from Moso International); Bus station in China (E, provided by Zhi-cheng Xue from Taohuajiang Bamboo Company).

in Colombia by guadua bamboo poles and concrete, this mode of construction with round poles is easier to replace structural components (Figure 6A). In China and South Africa, environmentally friendly parking lots were built by round bamboo poles and engineered bamboo materials respectively (Figures 6B, 6C). In the latter, photovoltaic solar panels were installed could make full use of clean energy. In addition, in some cities of Europe and China, also relevant institution uses engineered bamboo to make traffic signs or build bus station to replace high energy consumption materials such as steel and aluminum (Figures 6D, 6E). In some countries, the theft of aluminum road signs (for their scrap value) is common, with high cost and safety consequences. Bamboo has no scrap value, so is not attractive to thieves (Moso International, 2018). At present, although the uses of bamboo in transportation facilities are still in demonstration stage, the advocacy of low carbon transportation system in many countries will greatly promote potential use of biomaterials in transportation sector in the coming years.

2.2.5 *Outdoor landscape*

The applications of bamboo in outdoor landscape mainly revolves themes of “nature” or “fun”. Art Museum in Phoenix Valley located in Beijing, China (Figure 7A) was opened in August, 2018. Walkways built by round bamboo poles linked the undulations of roofs with natural mountains, blurred the boundaries between nature and man-made, while interpreting Chinese traditional culture with contemporary architectural techniques (dEEP Architects, 2018). The Panda Pavilion of Shenyang Forest Zoo, China, completed in 2017, has shifted its landscape design from artificial to natural for pandas in which visitors can also get close to nature (Figure 7B). The application of engineered bamboo materials for urban landscape mainly focus on outdoor footpaths and city furniture. Outdoor bamboo decking with high corrosion resistance, has a service

A

B

C

D

Figure 7. Bamboo used in outdoor landscape.

Photos source: photo of Art Museum in Phoenix Valley (A) and photo of Panda Pavilion of Shenyang Forest Zoo (B) are provided by Wei Cai from Zhujing Bamboo Company; Photo of Civic Service Centre in Xiong'an, China (C) and photo of public bamboo chairs in Wuhan, China (D) are provided by Zhicheng Xue from Taohuajiang Bamboo Company.

life of more than 25 years[6] and plays an important role in urban landscape. Figure 7C shows an outdoor square with bamboo flooring of Civic Service Centre in Xiong'an, China, which is a pilot city that China intents to build with the most advanced concepts and international standards for its urban design. Figure 7D shows public bamboo chairs in Wuhan, China, which create rest spaces for citizens with natural material.

2.2.6 *Eco-tourism practices in rural area*

Bamboo has a great potential for eco-tourism industry in many countries. China is a pioneer who has successfully used bamboo as a construction material in rural areas. For example, a youth hotel was built in Niubei Mountain, Sichuan, China in 2015 (Figure 8A) for which architect used engineered bamboo scrimber as the main roof structural material. The project not only provides a base for tourists in distress when they travel in mountains, but also provides long-term services for the elder and left-behind children in the village. Another inspired story happened in Baoxi, of Zhejiang, China in 2016, where18 bamboo buildings were inaugurated at the first International Bamboo Architecture Biennale. Within a short period of four years since 2012, a total of 13 international artists and architects used bamboo with other local materials such as rammed earth, stone and ceramics to create a myth of a small mountain village with abundant bamboo resources. Baoxi is now a cultural tourism village featuring bamboo architecture, that greatly improves the incomes of local people. It is estimated that an ordinary family hotel can earn at least $300,000[7] per year. More and more young people are willing to return from big cities to local employment and entrepreneurship that provides a dynamic and sustainable exploration path for future rural development (Figure 8B). Worth noting, the Chinese government released

6. Information provided by Arjan van der Vegte from Moso International
7. Information provided by Ge Qiantao, the Curator of the first International Bamboo Architecture Biennale

Figure 8. Bamboo used in eco-tourism practices in rural area.

Photos source: Youth Hotel in Niubei Mountain, China (A, provided by Hui Wu from Hongyazhuyuan Science and Technology Company); One of bamboo architectures at the first International Bamboo Architecture Biennale in Zhejiang, China (B, provided by Rongfu He from Selution Company); A boutique hotel in Zhejiang, China (C, source: http://loftcn.com/archives/82619.html (retrieve on 30th November, 2018)); Public space in one village of Anhui, China (D, provided by Cai Wei from Zhujing Bamboo Company).

a national policy of "Rural Revitalization" in 2017 which brings more opportunities for bamboo. Since then, a group of new eco-tourism bamboo construction practices boomed in the market. The following two projects are located in Zhejiang (Figure 8C) and Anhui (Figure 8D) provinces of China respectively where bamboo is a local material everywhere. Owing to the introduction of bamboo in the design, the former case successfully reformed a restaurant to a boutique hotel which attract more tourists to go there, and the latter achieved the organic renewal of village to create a new public space for local people.

2.2.7 *High-end custom-made bamboo products*

In recent years, the interest of architects or designers in natural materials has increased, but the price of high-quality hardwood continues to rise. Bamboo becomes a new material for high-end custom-made products.

The indoor space with round bamboo poles or bamboo weaving is a fashion way of designers. Due to the shape and characteristic of round bamboo poles or knitted bamboo products, the experiencer's vision can freely shuttle in the space, breaking the usual sense of closure for traditional architectural space. For example, a "Japanese Cuisine" restaurant in Hong Kong, China (Figure 9A) delicately used small diameter bamboo poles to divide the space, creating the atmosphere of traditional Japanese courtyard back to the 17th century. Another Thai restaurant in Taiwan, China (Figure 9B) used woven bamboo curtains to blur the boundary between restaurant and shopping corridors to create a lively Thai market. In addition, architects also applied bamboo weaving techniques to office buildings (such as Headquarters of Honeycomb Company in Beijing, China) (Figure 8C) and museums (such as Folk Art Museum of Chinese Academy of Art in Hangzhou, China) (Figure 9D).

Figure 9. High-end custom-made bamboo products (round bamboo poles or bamboo weaving products).

Photos: "Japanese Cuisine" restaurant in Hong Kong, China (A, source: https://www.cool-de.com/thread-1923686-1-1.html (retrieve on 30th November, 2018)); Thai restaurant in Taiwan, China (B, source: https://www.archdaily.cn/cn/899451/tai-wan-jie-dao-jian-tai-shi-can-ting-bo-cheng-she-ji (retrieve on 30th November, 2018)); Headquarters of Honeycomb Company in Beijing, China (C, source: https://house.focus.cn/zixun/b2a6c84362d6fff1.html (retrieve on 30th November, 2018)); Folk Art Museum of Chinese Academy of Art in Hangzhou, China (D, provided by Qiantao Ge from Shanghai Arts and Spring Company).

High-end custom-made engineered bamboo products are mainly used as interior decorative materials in international and domestic markets. Several suppliers, such as Moso International[8] from Netherlands, Lamboo[9] from the USA, Hangzhou Dasso Technology Company (Dasso)[10], Hunan Taohuajiang Bamboo Technology Company (Taohuajiang)[11] and Greenzu Bamboo Company (Greenzu)[12] from China and etc. provide innovative bamboo solutions for global customers. For examples, China Fuzhou Strait Culture and Art Center (Figures 10A,10B) was completed in 2018. In this project, hundreds of thousands of square meters of engineered bamboo materials were used as flooring, wall panels or grilles which needs to meet a series of strict requirements such as sound-proof, fire-proof and corrosion-resistance. In Europe, the largest shopping center in Italy, the CityLife Shopping District (Figures 10C,10D) was designed by Zaha Hadid Architects. The defining design factor of the shopping complex is the extensive and exquisite use of bamboo – the material flows from the flooring and extends into columns that eventually reach the ceiling in one swift wave-like motion. The bamboo was curved into shape using resins under high pressure and engineered blocks of the bamboo were carved into ribs by a 5-axis Computer Numerical Control (CNC) milling machine, then hand finished to create the totally smooth

8. https://www.moso.eu
9. https://www.lamboo.us
10. http://www.dassogroup.com
11. http://www.chinathj.cn
12. http://www.greezu.com

Figure 10. High-end custom-made engineered bamboo products.

Photos source: photos of China Fuzhou Strait Culture and Art Center (A and B) are provided by Tang Gangyi from Dasso; photos of CityLife Shopping District (C and D) are provided by José Luken from Moso International; photo of Li Zijian Art Museum (E) is provided by Zhicheng Xue from Taohuajiang; Photo of Hengsheng Management College in Hong Kong, China (F), photo of Bama Hotel in Guangxi, China (G) and photo of Jing'an Shangri-La in Shanghai, China (H) are provided by Eric Xiong from Greenzu.

and fluid interior effect (Harriet Thorpe, 2017). Except for the above two examples, high-end custom-made engineered bamboo products are also widely used in museums (Figure 10E), offices, school buildings (Figure 10F), hotels (Figure 10G), restaurants (Figure 10H) and etc.

3 CHALLENGES AND OPPORTUNITIES OF GLOBAL BAMBOO CONSTRUCTION INDUSTRY

In recent years, the global bamboo construction industry has made remarkable achievements: hundreds of large commercial projects are built of bamboo, many of which are designed by world renown architects. However, most of these projects are mainly driven by some small and medium enterprises (SMEs) in different countries. Due to our scientific understanding of bamboo as a construction material lags far behind other construction materials, bamboo doesn't have obvious advantages compared with other materials such as timber, concrete and steel. Therefore, SMEs and/or companies engaged in the bamboo modern construction often feel great pressure to survive. To improve the uptake of bamboo materials in mainstream construction markets needs joint efforts not only from private sectors, but also from researchers, architects, engineers, designers, developers and policy-makers. Therefore, in the final section of the paper, the authors analyze a number of challenges we need to tackle and point out several opportunities to guide the further development of global bamboo construction industry.

3.1 *Increase funding for basic research on bamboo construction materials*

There are total 1,642 bamboo species in the world (Vorontsova et al, 2016), of which only around 100 bamboo species could be used for construction.[13] However, bamboo is one of the most complicated construction materials in the world, which is not only anisotropic but also has different properties from its root to its top as well as from interior layer to exterior one. Even for a particular species in the same growing environment, different round bamboo pole has different mechanical properties. Therefore, the basic research of mechanical properties on different bamboo species is critical to the design of round pole bamboo structures. However, current researches are mainly focus on few species, such as *Phyllostachys edulis* (Moso) and *Guadua angustifolia* (Guadua). There is lack of data for most of other species when we want to use them in the construction. In addition, we still have no idea about exact geographic distributions and areas of those bamboo species. In the future, with the development of INBAR's Global Assessment of Bamboo and Rattan Resources (GABAR) Flagship Programme, the outputs of basic information of bamboo resources will guide the development of bamboo construction sectors in different countries.

On another issue, connections for bamboo structures are still poorly understood (Liu et al., 2013) that is another barrier for the development of round pole bamboo structures. Firstly, we should develop non-destructive methods for determining the strength of culms. Although the International Standardisation Organization (ISO) published the first international standard ISO19624 (Bamboo structures – Structural grading of bamboo culms – Basic principles and procedures) in September, 2018, the wide promotion of these methods in different countries still need some time. Secondly, we need to develop a simple and effective way for connection. There are more than 10 different types of connections in the world, such as lash, bolts, cement with bolts, steel connections and etc., but none of them are widely applied around the world due to different disadvantages.

However, inter-disciplinary approaches are opening another door for the future of round pole bamboo structures. For example, researchers from the UK started to use innovative ways to get required information of a single round bamboo pole. It takes only two minutes to scan a pole by a simple scanner, by using all the geometric and mechanical properties could be introduced into Building Information Modelling (BIM) to guide further design. For connections, researchers from Colombia are using 3D printing technology to print out connections directly, which is expected to be promoted and used in different round pole bamboo structures in the coming years.

13. The number of bamboo species for construction is based on the knowledge and reasonable study of multiple key experts of INBAR TFC.

With the development of modern bamboo manufacturing technologies, engineered bamboo materials could break the limits of bamboo species used in round pole bamboo structures and it is much easier to control the quality of final products. Although basic and applied researches on different types of engineered bamboo materials still need to be explored in the future, it is much easier to get funding for the research of engineered bamboo structures compared to round pole bamboo structures. Some countries, such as China, the USA and several European countries are keen to promote structural use of engineered bamboo in buildings. Due to the lack of codes and standards, the main use of engineered bamboo materials still focuses on interior decorations currently. However, the structural use of engineered bamboo is an inevitable trend in the future. In addition, with the development of Cross Laminated Bamboo Timber (CLBT), bamboo could also be used to construct multi-storey or high-rise buildings.

3.2 *Speed up the development national and international standards*

The majority of bamboo structures around the world do not comply with design codes, while most bamboo products lack internationally recognized standards (Liu et al., 2013). The lack or absence of universally recognized bamboo standards in the bamboo construction sector is one of the major barriers to scale up structural use of engineered bamboo materials. In an attempt to address this challenge, INBAR is leading an international group of bamboo construction experts to develop standards in ISO Technical Committee (TC)165 -Timber Structures Working Group 12 - Structural Uses of Bamboo. In addition to the published ISO19624 introduced in previous chapters of this paper, the group is currently revising two existing standards: ISO22156 & ISO22157-1 and two new standards which are testing methods and products of engineered bamboo. Upon release, these two standards will greatly improve the competitiveness of bamboo as a construction material in the international market.

Another strategic work of INBAR is to strengthen the collaboration with national standardization authorities of different countries to promote the adoption of ISO standards or to help them to develop their own national standards. For example, Dr. Xu Qingfeng[14] and Dr. Li Haitao[15] are two Chinese experts in INBAR TFC, who are also nominated as national experts of China in ISO/TC165, are currently leading the development of total four association standards of engineered bamboo in China. Under the current course, after two or three years, the outputs of association standards could on the one hand be used in the development of international ISO standards, and on the other hand help to provide a new chapter of bamboo structure in the future revising version of Chinese national timber standard.

3.3 *Enhance capacity building for bamboo construction professionals*

The majority of higher education and vocational training institutions across the world do not provide training for building professionals, such as carpenters, engineers, architects or designers on bamboo (Liu et al. 2013). It is estimated that no more than 100 universities in the world have bamboo building courses for students, and independent training centers for bamboo construction professionals are much fewer. However, the increasing interest of architects and/or designers in natural materials has resulted in an increasing need for trainings of bamboo as a construction material in many countries. As an intergovernmental organization, INBAR receives lots of requests from its member countries and private sectors across the world every year to request INBAR to provide capacity building services or technical supports in bamboo construction to them. INBAR has been working with many partners around the world to carry out a series of bilateral or multi-lateral trainings from the early of 2000s, but the number of trainees is far behind the number of required practitioners in bamboo construction sector.

14. https://www.inbar.int/tf_qinfeng-xu/
15. https://www.inbar.int/tf_haitaoli/

Nevertheless, there are still some opportunities while the increasing demands of technology transfer between countries. In recent years, those countries with advanced bamboo industry has been increasing investments to speed up the capacity building of those countries with less developed bamboo industry. Taking one training programme sponsored by the Ministry of Commerce of the People's Republic of China (MOFCOM) for example, INBAR has worked with the International Center for Bamboo and Rattan (ICBR), a local research institute in China and have jointly trained more than 1400 participants from around 80 different countries from 2005 to 2018. The number of trainees increases every year. Besides, in response to a demand from the Ethiopian government, China will build in the near future the Sino-African Bamboo Center in Addis Ababa. The center is expected to greatly promote the development and capacity building of the bamboo industry in Africa continent.

3.4 *Awareness raising through multiple ways*

Bamboo is perceived as a poor-man's timber in many countries which is mainly due to people's impression on the poor natural durability of round bamboo pole. However, with the development of the technologies of preservation and treatment, round pole bamboo structures can be also be used for more than 30 years if the structures with proper maintenance. Moreover, if a structure could be designed with flexible connections, the life span of the structure will not be determined only by the durability of bamboo poles since components could easily be replaced. Besides, we need to provide more opportunities to public to get in touch with those fabulous modern bamboo houses. In facts, modern bamboo houses could meet all the requirements of modern life through proper design, no matter if it is round bamboo poles or engineered bamboo materials. Although many countries have enacted national policies to promote the development of bamboo construction industry in recent years, it will take some time for the general public to accept modern bamboo construction.

4 CONCLUSIONS

In recent years, the global bamboo construction industry has made a great progress, more and more commercial projects boomed in the market. Due to different economic level and society culture in different countries, bamboo is mainly used as a construction material in the following six aspects of applications: 1) Earthquake or natural disasters resistant houses; 2) Low cost social houses; 3) Round pole bamboo structures and engineered bamboo structures; 4) Small transportation facilities; 5) Outdoor landscape; 6) Eco-tourism practices in rural area; 7) High-end custom-made bamboo products.

However, there are still many challenges to the wide-scale promotion of bamboo in construction. To address challenges depicted in this paper it needs an international and interdisciplinary effort as well as a coordinated effort from researchers, architects, engineers, designers, developer, policy-makers as well as entrepreneurs. At present, INBAR's new platform – INBAR TFC which already successfully coordinates many bamboo construction activities involving international research institutes and commercial companies in the past five years. With the aim of acting as the world's premier information and knowledge center on structural use of bamboo, INBAR TFC is dedicated to work with all the partners around the world to promote the development of global modern bamboo construction industry.

ACKNOWLEGEMENT

The authors would like to appreciate the following bamboo construction experts to provide photos and detailed information to this paper: Ms. Verónica María Correa Giraldo, Founder & CEO of Kaltia & Bambuterra, Mexico; Mr. Luis Felipe, Head of Research and Development Department of Base Bahay Foundation, Philippines; Mr. Markus Roselieb, Founder of Chiangmai Life Architects and Construction, Thailand; Professor Huang Dongsheng, Nanjing Forestry

University, China; Mr. Eric Xiong, General Manager of Greenzu Bamboo Company, China; Mr. Jeran Hammann, Executive Vice President of Lamboo, the USA; Mr. Weiren Zeng, Director of Anji Bamboo Research and Design Center, China; Mr. José Luken, Head of Marketing of Moso International, Netherlands; Mr. Arjan van der Vegte, R&D Manager of Moso International, Netherlands; Mr. Zhicheng Xue, CEO of Hunan Taohuajiang Bamboo Technology Co. Ltd, China; Mr. Wei Cai, CEO of Zhujing Bamboo Company, China; Mr. Zhong Wang, CEO of Hongyazhuyuan Science and Technology Co., Ltd, China; Ms. Hui Wu, Marketing Manager of Hongyazhuyuan Science and Technology Co., Ltd, China; Mr. He Rongfu, Manager of the Selution Company, China; Mr. Hai Lin, CEO of Hangzhou Dasso Technology Co., Ltd, China; Mr. Gangyi Tang, Marketing Manager of Hangzhou Dasso Technology Co., Ltd, China; Mr. Qingwei Hu, Vice-Manager of Flooring of Hangzhou Dasso Technology Co., Ltd, China; Mr. Qiantao Ge, CEO of Shanghai Arts and Spring Co.,Ltd, China. We are also grateful to Mr. Xavier de Jauréguiberry, the Photographer of Architecture from France to provide two photos to the paper.

In addition, we are grateful to all the other speakers of ICBS2018 and all the other participants who attended the conference. However, due to limited chapter, the full name list will not be listed here.

At last, we are grateful to the supports from the China Academy of Engineering Consulting Project "Research on Development Strategy and Key Technologies of Bamboo Construction Sector in China towards 2035" (No. 2018-ZCQ-06).

REFERENCES

BASE, 2018. https://www.base-builds.com, (retrieve on 30th November, 2018).

Sharma, B., Gatóo, A., Bock, M., and Ramage, M, 2015. "Engineered bamboo for structural applications." Construction and Building Materials, Volume 81, 15 Apr. 2015.

C&E, 2018. www.ceingenierie.fr/en/projet/bamboo-contemplation-exhibition-pavilion-in-arles-13/, (retrieve on 2nd December, 2018).

dEEP Architects, 2018. https://www.archdaily.cn/cn/902902/feng-huang-gu-shan-ding-yi-zhu-guan-deep-architects, (retrieve on 30th November, 2018).

Designboom, 2018. https://www.designboom.com/architecture/chiangmai-life-architects-bamboo-sports-hall-panyaden-international-school-thailand-08-09-2017//, (retrieve on 2nd December, 2018).

Gatóo, A., Sharma, B., Bock, M., Mulligan, H. and Ramage, M.H., 2014. Sustainable Structures: Bamboo Standards and Building Codes. Proceedings of the Institution of Civil Engineers: Engineering Sustainability, 167 (5), pp. 189–196.

Harriet Thorpe, 2017. Zaha Hadid Architects' CityLife Shopping District opens in Milan.

IBUKU, 2011. http://ibuku.com/pemulung-housing (retrieve on 30th November, 2018).

INBAR, 2018. Trade Overview 2016: Bamboo and Rattan Products in the International Market. International Organization for Bamboo and Rattan (INBAR).

Jiang Z. H., 2007. Bamboo and Rattan in the World, China. Forestry Publishing House.

Liu, K.W., Frith, O. B, 2013, An Overview of World Bamboo Architecture: Trends and Challenges. World Architecture, 2013 (12).

Mary Roach, 1996. The Bamboo Solution – Tough as steel, sturdier than concrete, full-size in a year. Discover Magazine. http://discovermagazine.com/1996/jun/thebamboosolutio784 (retrieve on 30th November, 2018).

Moso International, 2018. https://www.moso.eu/en/references/bamboo-traffic-sign-coevorden (retrieve on 30th November, 2018).

Vorontsova M.S., Clark L.G., Dransfield J., Govaerts R., Wilkinson T., Baker W.J., 2016. WorldChecklist of Bamboos and Rattans. International Organization for Bamboo and Rattan (INBAR).

S. Kaminski, A. Lawrence, D. Trujillo, 2016. Design Guide for Engineered Bahareque Housing.International Organization for Bamboo and Rattan (INBAR).

VHC, 2018. https://hogardecristo.org.ec (retrieve on 30th November, 2018).

VOA News, 2018. https://www.voanews.com/a/indonesians-discover-bamboo-and-wood-beat-concrete-and-steel/4524112.html (retrieve on 30th November, 2018).

Modern Engineered Bamboo Structures – Xiao, Li & Liu (eds)
© 2020 Taylor & Francis Group, London, ISBN 978-1-138-35185-1

INBAR Construction Task Force – An explorative way for development in the bamboo construction sector

K.W. Liu
INBAR, International Bamboo and Rattan Organization, China

D. Trujillo
Coventry University, UK

K. Harries
University of Pittsburgh, USA

M.C. Laverde
Studio Cardenas Conscious Design, Italy

R. Lorenzo
University College London, UK

O. Frith
International Rice Research Institute, The Philippines; formally (2007-2017) with INBAR

ABSTRACT: The INBAR Construction Task Force (TFC), facilitated by INBAR since 2013 and officially established in 2014, helps to coordinate activities of international research institutes and commercial companies interested in structural uses of bamboo. Prior to the task force's initiation, there was little coordination or communication between individual research teams or commercial companies working on this subject. Currently, the TFC consists of a core group of 28 experts from 18 countries, aiming to serve as the world's premier information and knowledge repository on structural uses of bamboo. In the past four years, the TFC has contributed many achievements including: standardisation work for structural uses of bamboo; high quality peer-reviewed publications; international projects; consultancy services; international conferences; and capacity-building for the bamboo construction professional.

1 BACKGROUND

Since the early 2000s, INBAR has served as the liaison organization to the Timber Structures Technical Committee of the International Organization for Standardization (ISO/TC165) and successfully led an international expert group in the development of two international standards (ISO 22156 and ISO22157-1) and one technical report ISO/TR 22157-2 in 2004. This work has had significant impact globally, with a number of INBAR member countries, such as India, Ecuador, Peru and Colombia, subsequently developing chapters on bamboo in their respective national building codes which reference these ISO standards.

Despite the impact of this initial work, and without funding for its continuation, INBAR was unable to sustain further development of international standards development in the years since then. The standards published in 2004 were in many respects "version zero" standards; they identified needs and served as placeholders upon which revisions could be based. The 2004 ISO standards, in particular, lacked guidance on crucial issues such as connection design and strength grading. Nonetheless, despite major advances taking place in the field of bamboo construction in the subsequent decade, no new or revised ISO standards for structural uses of bamboo were developed. Additionally, engineered structural bamboo products, which were relatively new in the early 2000s, have gained commercial importance but are, are not covered by existing international standards.

It is noted that since the first ISO bamboo standards were developed there has been some development and revision of national standards. Most significantly, in Colombia, the publication (and ongoing revision) of NSR-10 (AIS 2010) and the inclusion of composite bamboo "lumber" into ASTM D5456 in 2010.

Recognising the gap in standards development, in 2013, with new core funding from China, INBAR initiated new research on structural uses of bamboo through a strength grading project with Coventry University in the United Kingdom and partners in Colombia and Ecuador. The project was able to successfully obtain permission from ISO/TC165 to include a new Working Group 12 (WG12) on structural uses of bamboo. This working group has the mandate to amend existing ISO standards as well as to propose new ones. INBAR has served as the Convenor for this working group since its inception in 2014.

In addition to the project, INBAR had also successfully organised an informal network of about 15 research institutions and a few commercial companies that were researching structural uses of bamboo. This group met at the University of Cambridge in 2013 and at Coventry University in 2014. The process indicated that there was a growing international community of researchers (in Canada, China, Colombia, India, Indonesia, Malaysia, Mexico, USA and the UK) working on bamboo-based construction. Prior to the two meetings in the UK, however, there was little coordination or communication between individual research teams or commercial companies.

Despite the advances made by the China Government-funded project and INBAR's efforts to build informal networks, it was clear that in order to avoid repeating the experiences of post-2004, INBAR needed to put in place structures that would leverage different stakeholders working in this area to work together and pool resources. Therefore, INBAR proposed to set up a new task force to enhance its network and global reach on structural applications of bamboo.

The initial INBAR Construction Task Force (TFC), consisting of 11 experts from Canada, China, Ethiopia, India, Nepal, Peru, the UK and USA, was formally established in 2014. Four years later, the core group of experts has increased to 28 individuals in 18 countries representing a truly global reach (Figure 1).

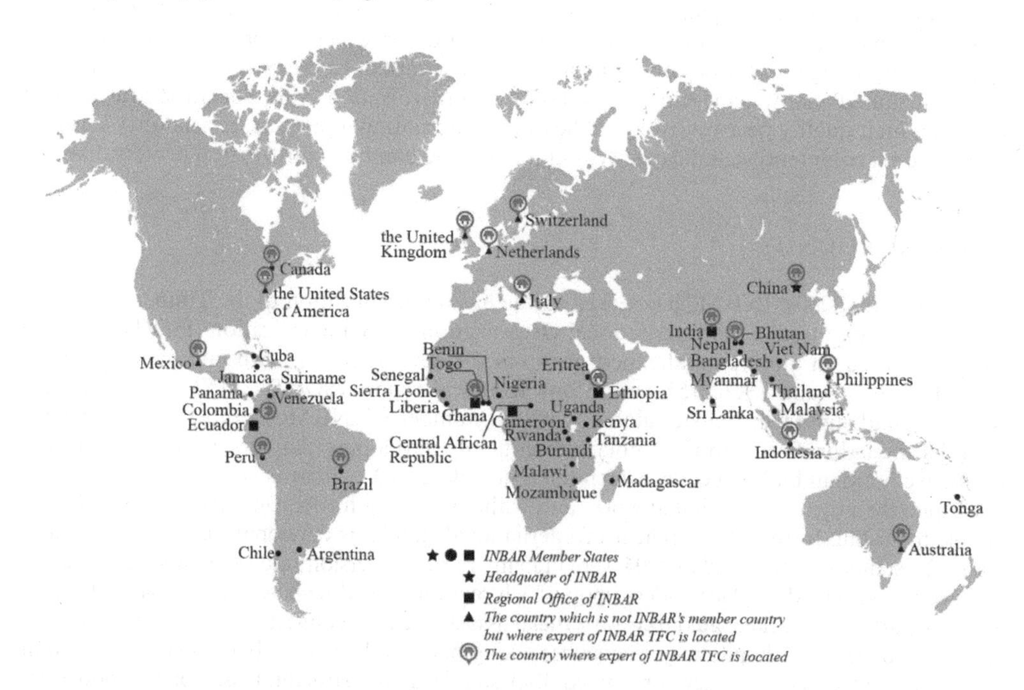

Figure 1. Distribution of INBAR TFC Experts.

2 INTRODUCTION

2.1 *Objective of INBAR TFC*

The primary objective of INBAR TFC is to act as the world's premier information and knowledge repository on structural uses of bamboo. The specific objectives are as follows: 1) help drive and refine development of new international standards on structural uses of bamboo, as well as help to review and update existing international standards in this area; 2) support global coordination and knowledge dissemination on sustainable bamboo construction; 3) facilitate the development of socio-economically appropriate methodologies for designing and constructing sustainable bamboo housing; 4) contribute towards capacity-building of construction sector stakeholders in sustainable bamboo housing; and, 5) raise awareness and advocate for bamboo construction being mainstreamed in national housing policies and regulations.

2.2 *The structure of INBAR TFC*

INBAR TFC has a management team consisting of a TFC Manager and a TFC Chair; the latter is a technical expert but not on INBAR staff. This management team is responsible for coordinating the task force. Both the Manager and the Chair are members of the ISO TC 165 WG12 and are responsible for ensuring that the actions of the TFC are integrated into ISO's international standard development. At present, Ms. Kewei Liu, the Global Bamboo Construction Programme Coordinator of INBAR is the Manager of INBAR TFC and David Trujillo, Assistant Professor at Coventry University in the UK is the Chair of the group. At present, the TFC experts are as follows:

Denamo Addissie: Addis Ababa University, Ethiopia
Nripal Adhikary: ABARI, Nepal
Yann Barnet: San Martin University, Peru
Ramesh Chaturvedi: Indian Institute of Technology (retired), India
Juan Francisco Correal Daza: University of the Andes, Colombia
Edwin Zea Escamilla: University of Zurich, Switzerland
Romildo D. Toledo Filho: Federal University of Rio de Janeiro, Brazil
Verónica María Correa Giraldo: KALTIA and BAMBUTERR, Mexico
Mateo Gutierrez Gonzalez: University of Queensland, Australia
Kent Harries: University of Pittsburgh, USA
Sebastian Kaminski: ARUP, UK
Sanjeev Kappe: Konkan Bamboo & Cane Development Center, India
Emmanuel Appiah Kubi: CSIR-Forestry Research Institute, Ghana
Mauricio Cardenas Laverde: Studio Cardenas Conscious Design, Italy
Andrew Lawrence: ARUP, UK
Haitao Li: Nanjing Forestry University, China
Luis Felipe Lopez: Base Bahay Foundation, Philippines
Rodolfo Lorenzo: University College London, UK
Michael Ramage: University of Cambridge, UK
Hector Archila Santos: Amphibia BASE Ltd, UK
Bhavna Sharma: University of Bath, UK
Greg Smith: University of British Columbia, Canada
Martin Tam: Able Mart Limited, Hong Kong, China
Caori Patricia Takeuchi Tam: National University, Colombia
Arjan Van Der Vegte: Moso International, The Netherlands
Andry Widyowijatnoko: Bandung Institute of Technology, Indonesia
Qingfeng Xu: Shanghai Research Institute of Building Sciences, China
Biographical information of all experts is provided on INBAR's website.

2.3 *Membership*

2.3.1 *The principles and the benefits of membership*

Membership in INBAR TFC is on a voluntary basis. Benefits of TFC membership are as follows: 1) experts are first option candidates to take on short consultancies for INBAR on bamboo construction related topics; 2) experts may leverage their membership when applying for grants; 3) experts publish their profile and links to their latest work through INBAR's network.

2.3.2 *How to become a member of INBAR TFC?*

INBAR TFC is open to new members, who can substantively add to the expertise of the group. To be admitted as a new expert to the INBAR TFC. Information on the application process is available from the TFC Manager - Ms. Liu Kewei - kwliu@inbar.int.

2.4 *Communications*

The TFC communicates primarily via email. Virtual meetings are organized for small groups of experts to discuss specific projects. The TFC meets once per year to discuss new standards and the amendment of existing ones, as well as to share research results and findings and agree on an annual action plan for the task force.

3 WHAT IS INBAR TFC DOING?

At present, INBAR TFC mainly focuses on the following works on bamboo construction related topics: 1) standardisation for structural uses of bamboo; 2) production of high quality publications; 3) international projects; 4)consultancy services; 5) international conferences; and, 6) capacity-building for bamboo construction professionals.

3.1 *Standardization work for structural uses of bamboo*

3.1.1 *International standardization work*

Since INBAR officially established TFC in 2014, ISO TC/165 WG12 consists of a group of bamboo construction experts mainly from INBAR TFC. Many TFC members have been nominated representatives of or "advisors" to their TC165 national standardisation bodies. As introduced in the background of this paper, INBAR has engaged in the international standardization work since the early 2000s. Through the end of January 2019, ISO published four international standards and one technical report related to bamboo construction as follows:

ISO22156:2004 Bamboo – Structural Design;
ISO22157-1:2004 Bamboo – Determination of Physical and Mechanical Properties – Part 1: Requirements; (Withdrawn)
ISO/TR 22157-2:2004 Bamboo – Determination of Physical and Mechanical Properties – Part 2: Laboratory Manual; (Withdrawn)
ISO19624:2018: Bamboo Structures – Grading of Bamboo Culm – Basic Principles and Properties;
ISO22157:2019: Bamboo Structures – Determination of physical and mechanical properties of bamboo culms – Test methods

ISO22156, ISO22157-1 and ISO/TR 22157-2 were published in 2004. In January 2019, ISO 22157:2019 was published while ISO22157-1 and ISO/TR22157-2 were withdrawn. Led by David Trujillo, the revised Standard ISO 22157 adds two additional test methods not included in 2004 and revises a number of other test methods with the intent of improving the utility – and therefore hopefully the adoption – of the methods and Standard. It is the intent of WG12 to withdraw ISO 22157-2. This "Laboratory Manual" would provide better utility as a curated 'living document' maintained by INBAR to support the use of ISO 22157:2019.

In September 2018, ISO published another new standard ISO19624:2018 which deals with the structural grading of bamboo culms for construction. The new standard is the final output of an INBAR-funded project, "Strength grading of bamboo", led by David Trujillo from 2013 to 2016. Trujillo then led the ISO 19624 standard development effort from 2014 to 2018. This was the first ISO Standard developed with the support of the INBAR TFC. The aim of this standard is to provide the framework for any national or locally adopted grading process adopted anywhere in the world. It identifies criteria that should be considered in a visual grading protocol, and outlines the basis on which declared capacity or strength values would be arrived at.

In addition to the published ISO 19624 and ISO 22157, WG12 is currently revising another existing ISO standard (ISO 22156) and developing two new standards (NWIP23478 and another standard about engineered bamboo products yet not formally proposed to ISO) as follows:

- Revision of ISO 22156:2004. Led by Kent Harries, a complete revision of ISO 22156 is underway. This revision will provide both capacity and stress-driven approaches to design and provide means of design or acceptance of details, particularly for connections. The revised document builds upon the considerable research and development conducted in the last 20 years and better integrates ISO 22157, ISO 19624 and other modern ISO standards. The work is presently underway at the Committee Document stage: CD22156: Bamboo Structures — Bamboo Structural Design
- NWIP23478: Bamboo Structures — Engineered Bamboo Products — Test Methods for Determination of Physical and Mechanical Properties. This new test methods Standard, led by Arjan van der Vegte and Bhavna Sharma, is a landmark for the development of international standardization work for engineered bamboo materials. Since the 1990s, industrially produced engineered bamboo products like bamboo flooring have been developed and introduced to the international market. Recently, engineered bamboo is being used increasingly for external features of buildings, as well as interior finishes. This has boosted the demand for engineered bamboo products (van der Vegte, 2017). This new project will develop test methods mainly for two types of engineered bamboo products: glue laminated bamboo and bamboo scrimber.

In order to make clear that the scope of standards what ISO TC165 is developing is different from ISO TC296 (Bamboo and Rattan Technical Committee, which was established in 2015 with its Secretariat in China), the title of the standards to be developed in TC165 will begin with "Bamboo Structures".

3.1.2 *National standardization work*

In addition to the work on international standardization, INBAR also works very closely with national standardization authorities in different countries, like China, Ecuador, India, Ethiopia and Kenya. For example, INBAR has close collaboration with the China Southwest Architectural Design and Research Institute (CSCEC) which mirrors ISO/TC165 and the China Association for Engineering Construction Standardisation committees responsible for bamboo and timber standards. CSCEC helps to nominate Chinese experts to participate in the development of international standards of bamboo construction in ISO/TC165 and is the management organization supervising the development of association standards for bamboo construction. Since China revised its standards policy in 2015, the status of association standards has been elevated while it is difficult to apply for any new national standards in the future. Currently, two Chinese members of INBAR TFC, Xu Qingfeng and Li Haitao, are leading the development of four association standards for engineered bamboo in China. As CSCEC is very familiar with the international standardization work in ISO as well as for the association standards in China, it is possible a new chapter on bamboo structures will be developed for future revisions of the Chinese national timber standard.

3.2 *International projects*

Although the membership and all the work of INBAR TFC is on a voluntary basis, the relationship between members have been enhanced greatly since the group was established. In the past four years, several international projects were successfully implemented with the cooperation of the INBAR TFC.

3.2.1 *Bamboo in the urban environment*

Led by the University of Pittsburgh and Coventry University with INBAR as a partner and funded jointly by the US Department of State and UK Council, this project brought together leading experts in various streams of bamboo research to carry out extensive, cutting edge analysis and testing to further enable the safe use of bamboo in urban centers. The resulting collaborative research has significant technical and social relevance through its potential to reduce the cost and environmental impact of safe housing for a significant proportion of the world's population. This work addressed the global grand challenges of urbanization and resilience in the face of natural hazards and climate change through facilitating the use of a renewable 'green' material.

The project funded three international symposia (Winnipeg 2015, Pittsburgh 2016 and Bogor (Indonesia) 2017) reaching (and supporting) a total of 83 registered participants representing 18 countries, 25 universities, 14 companies, 5 inter/governmental agencies and 3 NGOs. The attendees ratified the Pittsburgh Declaration in 2016 and reconfirmed this in 2017. The project also supported three international graduate student research exchanges. At least 15 journal articles have been developed from the collaborations initiated as a result of this project. A number of research proposals, including successful US National Science Foundation funding also resulted.

3.2.2 *The BIM bamboo project*

Led by University College London (UCL) and funded by the UK Engineering and Physical Sciences Research Council (EPSRC) this project is focused on developing new digital design and fabrication workflows for bamboo poles following the principles of Building Information Modelling. INBAR's support enabled a field trip to China to trial these processes and introduce stakeholders to the potential of these technologies to stimulate growth in the bamboo construction sector and reduce the negative environmental effects of intensive manufacturing. The close links developed following this field trip led to a successful application to the UK-China Joint Research and Innovation Partnership Fund (British Council/CSC) to further develop this research during a six-month PhD research placement in Nanjing Forestry University.

3.2.3 *INBAR Garden Pavilion*

For the upcoming 2019 Beijing Horticultural Expo, INBAR invited architect Mauricio Cardenas Laverde to work with a local engineering company in China to design and construct an innovative INBAR Garden-Pavilion having an area of 3600 square meters. This project will provide a great opportunity for INBAR to showcase contemporary uses of the traditional natural material "Bamboo". The architectural concept, from which the project has been developed through all the following phases, is that of melting together architecture, engineering and landscape in order to express bamboo to its maximum. This idea presented by the Colombian architect based in Milan Italy in the architectural Design Concept and Design Development has generated a fruitful international cooperation between experts from different professional fields and backgrounds following the research and innovation approach of INBAR and its Task Force.

3.3 *Production of high quality publications*

3.3.1 *Technical report or working report*

Since its establishment in 2014, INBAR TFC has published a series of technical or working reports about bamboo construction. Electronic versions are available on INBAR's website and can be downloaded for free.

Figure. 2. Bamboo Test-kit-in-a-Back Pack (Harries et al. 2016).

3.3.2 *Bamboo Test-kit-in-a-Back Pack (Figure 2)*
This digital publication series illustrates the fabrication and use the "Test-kit-in-a-Back Pack" for rapid field assessment of bamboo mechanical and material properties using four ISO 22157 test methods (Glucksman and Harries 2015). It includes two parts: the Technical Report describes the fabrication of and provides reference for the use of the kit. the User's Manual, presently available in six languages (English, Spanish, Portuguese, Chinese, Haitian Creole and French), is a simple illustrated guide to performing the tests.

3.3.3 *Grading of bamboo (Figure 3a)*
This INBAR Working Paper (INBAR WP) presents research into potential grading methodologies for one species of bamboo – *Guadua angustifolia* Kunth – and recommends criteria for both visual g and machine grading (Trujillo, 2016). The report also represents critical background material for the development of ISO 19624. For visual grading, the diameter of bamboo culms is deemed to be an important consideration when grading for flexural capacity. Wall thickness is considered to be critical to shear and tension perpendicular capacities. For machine grading, three main properties were found to be significant: flexural stiffness is important for the design of beams; slender struts and portal frames; and linear mass is adopted to infer density. The INBAR WP suggests that additional properties that may be critical to design for other applications should be further researched including: shear strength or tensile strength perpendicular to fibers; both of which are important to the process of connection design. Implemented effectively, the grading methodologies presented in the report have the potential to enhance the supply of bamboo and deliver positive engineers, and consumers. The English e-version is available on INBAR's website.

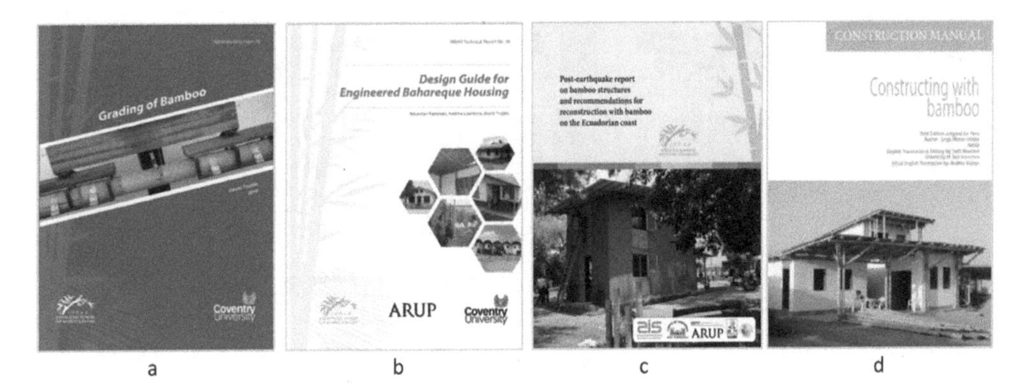

Figure. 3. Published INBAR TFC Publications.
Photos are from INBAR: Grading of Bamboo (a); Design Guide for Engineered Bahareque Housing (b); Post-earthquake report on bamboo structures and recommendations for reconstruction with bamboo on the Ecuadorian coast (c); Constructing with bamboo (d).

3.3.4 *Design Guide for Engineered Bahareque Housing (Figure. 3b)*

Engineered bahareque is a modified form of construction that takes traditional wattle-and-daub type housing and improves upon it, using modern materials, knowledge, and construction techniques (Kaminski et al. 2016). Engineered bahareque houses have successfully been constructed in various countries including: Costa Rica, Colombia, Nepal, Ecuador, El Salvador, and The Philippines. When properly designed and built, they have demonstrated their effectiveness as an affordable, hazard-resilient, safe and durable form of housing. This technical report is intended as a guide for both architects and engineers, addressing both conceptual and detailed design and construction of engineered bahareque housing in both developed and developing countries around the world. The report provides guidance on design for structural (wind and earthquake) loads. Typical construction details are also provided, along with guidance for quality control during construction. The English, Spanish and French e-versions are available on INBAR's website.

3.3.5 *Post-earthquake report on bamboo structures and recommendations for reconstruction with bamboo on the Ecuadorian coast (Figure 3c)*

This INBAR TP describes the potential for bamboo to be used more widely in Ecuador (and in other countries) to build low-cost housing (Drune et al. 2016). Topics covered include the high tolerance of bamboo to earthquake loads, how homes can be built from bamboo to make them earthquake-resistant, and how to ensure that homes built from bamboo have lasting durability. The report also covers the availability of bamboo supplies and how these might be improved. The English and Spanish e-versions are available on INBAR's website.

3.3.6 *Constructing with bamboo (Figure 5d)*

The translated third edition of this construction manual (Ubidia, 2015), first published in 2005, provides a basic, hands-on guide to using bamboo for construction: from choosing and treating the raw material, to putting together durable and safe structures. It draws heavily on experiences from the Andean region of Latin America, where bamboo is a vernacular construction material. The publication includes photographs and diagrams to illustrate best practices. The English and Spanish e-versions are available on INBAR's website.

3.3.7 *INBAR Construction Task Force newsletters (Figure 6)*

INBAR also published a series of INBAR Construction Task Force newsletters which aim to share the latest research, projects, events and publication information from all TFC members with INBAR's member states and all the stakeholders of bamboo construction.

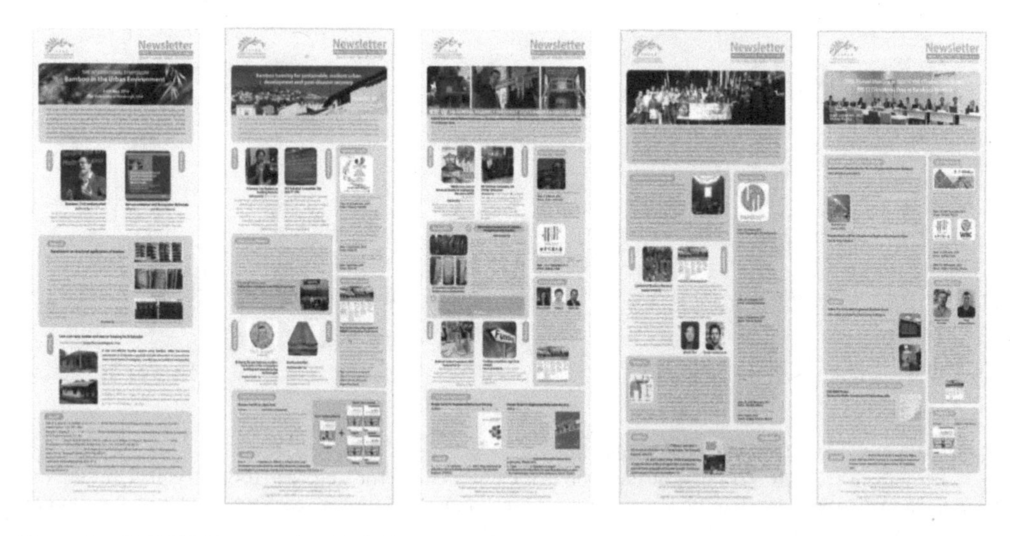

Figure 6. INBAR Construction Task Force newsletters.

3.4 *Consultancy services*

With the resources of bamboo construction experts around the world, INBAR TFC began offering consultancy services for third parties in 2017. For example, CRS (Catholic Relief Services) in India commissioned INBAR to conduct a study on bamboo shelters in the State of Odisha where frequent flooding, cyclones and monsoon rains cause widespread damage to shelters. The aim of the project is to understand the feasibility of using locally available bamboo for shelter construction, and to recommend strategies and approaches to improve bamboo shelters in both post-disaster and long-term development context. Sebastian Kaminski, an expert of INBAR TFC from Arup accepted the task and carried out a field visit. The assessment found that bamboo clearly plays an essential role in the shelter and ancillary structures throughout rural communities in the State. Unfortunately, current levels of knowledge along with the adoption of techniques that support optimal utilization of bamboo for shelter are low.

As a unique intergovernmental organization working on bamboo and rattan, INBAR receives many inquires; enquiries about construction are most frequent and include queries on methods of treatment, how to build houses using local species, and where to get training about bamboo construction, etc. Although the experts of the INBAR TFC do their best to provide the information requested, the most frequent inquiries should become the topics of the coming INBAR publications.

3.5 *International conferences*

Since 2015, INBAR TFC has been organizing many international conferences on bamboo construction around the world to promote the internal communication between TFC experts while disseminating the latest knowledge and information of bamboo construction to stakeholders as well as the public. The Bamboo in the Urban Environment project (see above) provided significant resources in this regard and helped to accelerate ISO Standards development efforts.

For disseminating the latest knowledge and information of bamboo construction, INBAR organized several international conferences around the world. For instance, in 2016, a side event, "Bamboo housing for sustainable, resilient urban development and post-disaster recovery" was organized by INBAR in Quito during Habitat III. Several TFC experts were invited to introduce their own experiences on building temporary bamboo shelters after natural disasters. During the World Bamboo and Rattan Congress 2018 (BARC2018) which was held in June, 2018 in Beijing, INBAR organized the Sustainable Bamboo Building Materials and Third International Conference on Modern Bamboo Structure (ICBS2018) with local partners in China. Almost half of TFC experts were present in Beijing to share the latest information of bamboo construction with more than 1000 participants from around the world.

An important effort in bamboo construction is to advocate and bring the material into mainstream construction. In April 2019, a number of TFC members will lead a special session at the American Society of Civil Engineers (ASCE) Structures Congress on bamboo materials and construction.

3.6 *Capacity-building for bamboo construction professionals*

Most of TFC experts are from global universities where they teach and train students from architecture, engineering and landscape departments every year. However, the majority of higher education and vocational training institutions across the world do not provide exposure to or training related to the use of bamboo materials for construction (Liu and Frith 2013). Therefore, professional and trades capacity-building for bamboo construction is an urgent need.

INBAR joined the effort with Prof. Xiao's team in China and trained more than 100 students from around the world in the "Village Bamboo Summer Programme" from 2017 to 2018. However, the number of trainees is limited due to limited financial support. The issue of adequate support cannot be understated: while students from most developed countries can generate financial support for such activities, few students from developing areas have access to the same resources. Therefore, we need to look for other ways to involve more students from around the world to get chances to interact with bamboo.

INBAR is planning to organize the International Bamboo Construction Competition (IBCC) in 2019 with the aim of providing an opportunity for college students of architecture, civil engineering, landscape and other bamboo construction related majors to get access to bamboo. The main theme of IBCC is to explore the potential use of existing bamboo construction materials in the market, for both round bamboo poles and various types of engineered bamboo. The competition topic(s) will be designed by TFC experts. Wherever participating students come from, they all will have chances to build bamboo houses once they participate in the competition when selected as finalists.

4 THE FUTURE WORK PLAN OF INBAR TFC

On the track of current development, the TFC will focus on the following aspects of the work about bamboo construction in the next few years:

- Promote the international standardization work step by step and promote the adoption of ISO standards by more countries;
- Strengthen cooperation with local standardization authorities in different countries to promote the development of national or regional standards;
- Carry out international projects of bamboo construction in Asia, Africa and South America.
- Publish more bamboo construction publications to meet the need of all stakeholders; and,
- Find sustained financial support to strengthen the capacity-building for bamboo construction professionals.

5 CONCLUSION

INBAR TFC is an explorative way to promote the development of a global bamboo construction industry. Although it is a new concept, a healthy group has grown up with many achievements on bamboo construction related topics in the past four years. Aiming to act as the world's premier information and knowledge repository on structural uses of bamboo, INBAR TFC sincerely invites bamboo construction experts from around the world to work with us.

REFERENCES

AIS, 2010, *Reglamento Colombiano de Construccion Sismo Resistente NSR-10*. Asociacion Colombiana de Ingeniería Sísmica.

ASTM International, 2018, *ASTM D5456 – 18 Standard Specification for Evaluation of Structural Composite Lumber Products*.

Drunen, N. Cangás, A. Rojas, S. & Kaminsky, S. 2016. Post-earthquake report on bamboo structures and recommendations for reconstruction with bamboo on the Ecuadorian coast, International Bamboo and Rattan Organization https://resource.inbar.int/download/showdownload.php?lang=cn&id=167873 (English e-version) https://resource.inbar.int/download/showdownload.php?lang=cn&id=167812 (Spanish e-version)

Harries, K. & Glucksman, R. 2016. Bamboo Test-kit-in-a-Back Pack, Part 1: Technical Report, International Bamboo and Rattan Organization https://resource.inbar.int/download/showdownload.php?lang=cn&id=167858 (English e-version)

Glucksman, B. & Harries, K.A. (2015) In-the-Field Test Methods for Bamboo – The test-kit-in-a-back-pack, *Proceedings 15th International Conference Non-conventional Materials and Technologies (NOCMAT 15)*, Winnipeg, Canada. August 2015.

Harries, K. & Glucksman, R., 2016, Bamboo Test-kit-in-a-Back Pack, Part 2: Users' Manual, International Bamboo and Rattan Organization https://resource.inbar.int/download/showdownload.php?lang=cn&id=167859 (English e-version) https://resource.inbar.int/download/showdownload.php?lang=cn&id=167860 (Chinese e-version) https://resource.inbar.int/download/showdownload.php?lang=cn&id=167861 (Spanish e-version) https://resource.inbar.int/download/showdownload.php?lang=cn&id=167862 (Portuguese e-version) https://resource.inbar.int/download/showdownload.php?lang=cn&id=167882 (French e-version, 2017)

ISO22156:2004, Bamboo — Structural Design, International Standardisation Organization.

ISO22157:2019, Bamboo Structures — Determination of Physical and Mechanical Properties of bamboo culms — Test methods.

ISO19624:2018, Bamboo structures — Grading of bamboo culm — Basic principles and properties, International Standardisation Organization.

Kaminski, S. Lawrence, A. & Trujillo, D. 2016. Design Guide for Engineered Bahareque Housing, International Bamboo and Rattan Organization https://resource.inbar.int/download/showdownload.php?lang=cn&id=167811 (English e-version) https://resource.inbar.int/download/showdownload.php?lang=cn&id=167914 (Spanish e-version, 2017) https://resource.inbar.int/download/showdownload.php?lang=cn&id=167914 (French e-version, 2017).

Liu, K.W. & Frith, O. B. 2013. An Overview of World Bamboo Architecture: Trends and Challenges. World Architecture, 2013 (12).

Trujillo, D. 2016. Grading of Bamboo, International Bamboo and Rattan Organization https://resource.inbar.int/download/showdownload.php?lang=cn&id=167810 (English e-version).

Ubidia, J. M. 2015. Constructing with Bamboo, International Bamboo and Rattan Organization https://resource.inbar.int/download/showdownload.php?lang=cn&id=167632 (Spanish e-version) https://resource.inbar.int/download/showdownload.php?lang=cn&id=167919 (English e-version, 2018).

Vegte, A.V.D. 2017. INBAR Construction Task Force Newsletter 2017Q2, International Bamboo and Rattan Organization https://resource.inbar.int/download/showdownload.php?lang=cn&id=167884.

Modern Engineered Bamboo Structures – Xiao, Li & Liu (eds)
© 2020 Taylor & Francis Group, London, ISBN 978-1-138-35185-1

A case study in bamboo construction education: The experience of the Sino-Italian "Bamboo Pavilion Contest – A contest for a pavilion in bamboo technology"

A. De Capua & M. Tornatora
Department of Architecture and Territory, Mediterranea University of Reggio Calabria, Reggio Calabria, Italy

C. Demartino & Z. Li
College of Civil Engineering, Nanjing Tech University, Nanjing, China

Y. Xiao
Zhejiang University-University of Illinois Institute (ZJUI), Zhejiang University, Haining, Zhejiang, China

ABSTRACT: This manuscript summarizes the experience of the Sino-Italian "Bamboo #Pavillon Contest_A contest for a Pavilion in Bamboo Technology" held between the Mediterranea University of Reggio Calabria (Italy) and the College of Civil Engineering of Nanjing Tech University (China) in 2018. The contest for the pavilion to be realized with bamboo technology was divided into three phases. The first one was aimed at identifying the best project idea among those presented from students of Mediterranea University of Reggio Calabria. The second phase was comprised of both the technical design and the construction of the bamboo pavilion by the students of Nanjing Tech University. The last phase consisted in the closing ceremony of the contest held at the Zhejiang University - University of Illinois at Urbana Champaign Institute. Particularly, the goal was the design of a temporary exhibition pavilion to be realized with bamboo structures, such as GluBam, an eco-sustainable building material also light and flexible but solid and resistant. The bamboo pavilion was realized and completed during the summer of 2018 and now it is a good example of a temporary structure made of engineered bamboo.

1 INTRODUCTION

The contest "Bamboo #Pavillon Contest_A contest for a Pavilion in Bamboo Technology" was held between the Mediterranea University of Reggio Calabria (Italy) and the College of Civil Engineering of Nanjing Tech University (China) in 2018.

The contest is a part of the joint actions provided by the Memorandum of Understanding (MOU) for Faculty and Student Mobility. It was signed by the two institutions in Nanjing on July 7th, 2017 (see Figure 1) for promoting between the two Universities both study and research relationships along with the exchange of students, PhD students and faculties. During the meeting and the signature of the MOU, the two Universities decided to start cooperation for realizing an innovative structure by joining the good knowledge of architecture from the Italian side with the good engineering knowledge of the Chinese side.

On the Italian side, Prof. Marina Tornatora developed a lot of research in order to deeply understand the meaning of Pavilion in architecture (Tornatora, 2017). As a matter of fact, in architecture, a pavilion has several meanings. The theme of the pavilion is proposed as the moment of materialization of a "thinking" idea: its nature is characterized by an ephemeral condition – given its limited life – and this becomes an opportunity for both continuous verification and experimentation of the "Limits" related to the project. This allows – from a design

Figure 1. Signature of the Memorandum of Understanding (MOU) occurred in Nanjing on July 7th, 2017. Picture on the left, from left to right: Dr. Zhi Li, Prof. Xiao Yan, Prof. Alberto De Capua, Dr. Cristoforo Demartino and Domenico Rositano. Picture on the right, from left to right: Dr. Zhi Li, Prof. Yan Xiao and Prof. Alberto De Capua.

point of view – the freedom to define special shapes in order to satisfy the functional requirements. Technological aspects have been treated by prof. Alberto De Capua expert on sustainable construction technologies (De Capua, 2015-2017).

On the Chinese side, the research group lead by prof. Yan Xiao is considered a leader in the use of engineered bamboo-based materials for structural applications (e.g., Xiao, 2009, Li et al., 2014, Xiao and Ma, 2014, Xiao et al. 2014, Wang et al., 2018). It uses the GluBam® technology invented by Prof. Xiao, a technology named by Popular Science magazine in its "Best of What's New in 2008" feature. The milestones in the use of bamboo for modern construction realized by his research group can be considered the demonstration of the California style house – constructed of bamboo material on Hunan University campus in Changsha, China – and the first modern bamboo truck-loaded roadway bridge in 2007 in Leiyang, China.

The structure of this paper is organized following the three phases of the contest. First, the contest for the students of Mediterranea University of Reggio Calabria (Section 2). In particular, the motivation of the contest and the requirements are introduced. The different proposals design and the winning one are presented. Then, the design and construction of the bamboo pavilion – done by the students of Nanjing Tech University – are presented (Section 3). Moreover, some information on the closing ceremony of the contest held at the Zhejiang University - University of Illinois at Urbana Champaign Institute (Section 4) is provided. Finally, some conclusions and perspectives are drawn (Section 5).

2 THE CONTEST

The Mediterranea University of Reggio Calabria (Italy) and the College of Civil Engineering of Nanjing Tech University (China), announced the "Bamboo #Pavillon Contest_A contest for a Pavilion in Bamboo Technology" on December 2017. The idea of a pavilion is related to the main features of this typology in the architecture. As a matter of fact, the pavilions are extraordinary for their temporary character (Piatkowska, 2013). The temporariness factor and the cost-effectiveness of pavilions' constructions are juxtaposed with high expectations due to the pavilion's prestige and the aesthetic expression. According to the circumstances often there are built objects that become architectural icons or unforgettable symbols and physical proclamation of the present.

The contest consisted on the choice of the best design idea among the presented ones. The participation to the contest is held for the students and Ph.D. students (or obtained within 5 years) who are regularly enrolled at the Mediterranea University of Reggio Calabria. The design project, for the exhibition pavilion, must respect the following requirements:

- Realized mainly with bamboo (raw bamboo, GluBam beams or panels) and designed to enhance the expressiveness of the material;

- The maximum size must be within a cube with side 5.4 m;
- Have a free exhibition space, also flexible and transformable.

The calendar of the contest was as follows: December 18th, 2017 – call for proposals; January - 12th, 2018 – deadline to send questions; January 30th, 2018 – deadline for receipt of project proposals. The award of the contest was the realization of the design idea and the participation to the Bamboo summer school at Nanjing Tech University.

The contest had two main objectives to stimulate students in innovative project activities. The first one consists in promoting the use of bamboo technologies, such as GluBam. The second one is related to the development of a discussion about the theme of the pavilion, that today is a hot topic in architecture. The ephemeral and multiform nature of the pavilion makes it privileged for spatial research and, therefore, a sensitive indicator of the profound changes in architecture and art. This character made the pavilion particularly suitable for a design competition that wants to be an opportunity for research and teaching together. In fact, the theme was proposed as a device of knowledge capable of materializing an idea that is still thought, expression of art and technique.

2.1 *Proposals*

10 groups of students (a total of 50 students) sent their design ideas to the contest. A summary of the design ideas is reported in Figure 2. Left-to-right then top-to-bottom in Figure 2, the proposals are:

1. [RE]frame - [Re]frame is an architectural design experiment referring to two main issues, namely: the notion of Rethinking the concept of the pavilion in the contemporaneity by Reusing and Reinterpreting the historically developed concepts, and the notion of frame as a tool towards the ephemeral character of the pavilion.
2. Blooming pavilion - Each perspective of the pavilion recalls a blooming flower shape, in the plan as well as at the eye-level perception. From the outside, the pavilion is glowing and fluctuating, thanks to the translucent fabric shell. Its petals overlap, in a play of light, shadows and transparencies. From the inside, the flower reveals, instead, the strength of its structure in GluBam technology.
3. BAM pavilion - Beyond Architecture Mutation. Space in a Convergence of Time is focusing to unfold several moments in time, able to emphasize the different qualities of conscious and unconscious experiences.
4. Lotus Amoenus - In China the Lotus flower is more than a natural beauty to be observed: behind its delicate petals, it hides a very deep spiritual meaning. The concept of the pavilion comes from this concept. The figure can be traced within a circle inscribed in a square.

Figure 2. The 10 design proposals of the contest.

5. Bamboo wave - The designer team decided to adopt 27 pieces of streamline 20-meter long wooden bamboo forming a fluctuating body. The pavilion is made of advanced fiber by hot gluing technology; it has the characteristics of high strength and high toughness.
6. Bamboo in the ephemeral space - recognizing in the theme of pavilion the intrinsic value of the ephemeral and the need not to consider it as an inhabited building, but a temple/laboratory. The idea of this project is configured as an open object in all its parts, playing with the transparency of the form and with the lack of distinction between the inside-outside of the exhibition space.
7. RIB Resisant Innovative Bamboo - The expositive pavilion Rib has been designed to valorize the structural technology of bamboo. The main elements are the arches (the "ribs"), two lamellar beams in bamboo with a rectangular section that are connected at the top and partially supported by a sequence of straight culms.
8. Bamboo#Pavilion - The project is proposed as a multisensory device that simultaneously involves sight, hearing, smell and touch. From a conceptual point of view, the scheme consists of three elements: the BASE, the YIN module (white) and the YANG module (black) which are characterized by four sub-modules of light (bamboo Ivory) and dark (carbonized bamboo) coloring.
9. The Cube and the Sphere - The project consists of a design for an expositional pavilion realized with bamboo technology. The basic idea is to combine the pure, symmetric volumes of the Cube and the Sphere into a single architectural object.
10. Comfortably Numb Pavilion - The name is from the well-known Pink Floyd's composition. The pavilion stems from a retelling of the peculiar relationship between the "irrationality" of the bamboo's growth process and the will of making the organic essence of this material rational, since it has the capacity of behaving as a structure generating spaces "domesticated" by the architectural gesture.

2.2 *The winning idea: the [RE]FRAME bamboo pavilion*

The work of the Committee[1] was not simple for the high-quality of all the design projects. The three winner design projects were in order: 1. [RE]frame; 8. Bamboo#Pavilion; 4. Lotus Amoenus. The groups were awarded during a ceremony held in Italy (Figure 3).

[Re]frame was designed by Blagoja Bajkovski, Cosimo Metastasio, Matteo Milano, Antonia Vadalà, Stefano Vitale. The intention of [Re] frame is the materialization of a flexible and adaptable pavilion to different contexts and places without imposing itself on the landscape, but at the same time capable of insinuating its roots in ever different terrains, with the

Figure 3. The award ceremony at the *Mediterranea* University of Reggio Calabria. From left to right: Prof. Gianfranco Neri, Prof. Francesco Carlo Morabito, Prof. Marina Tornatora, Prof. Alberto De Capua, Prof. Yan Xiao.

1. Prof. Gianfranco Neri, Prof. Yan Xiao, Prof. Alberto De Capua, Prof.ssa Marina Tornatora, Dott. Zhi Li, Dott. Cristoforo Demartino.

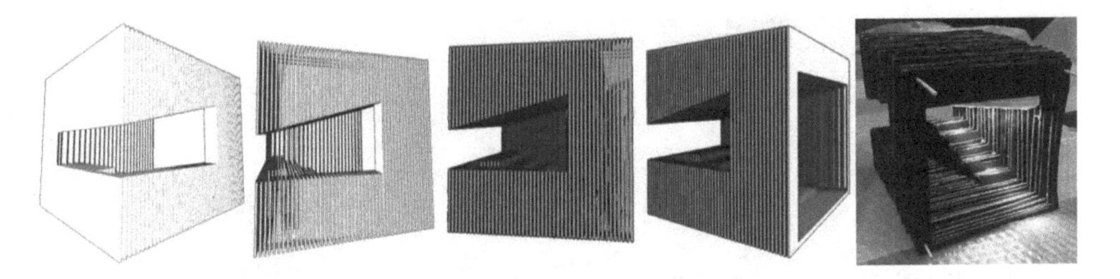

Figure 4. Mutable perception of the pavilion. Preliminary studies on a simplified model (right).

maximum integration (see Figure 4). Reuse and re-adaptation are the strongest features of the pavilion, which lends itself to different needs and can perform different functions in different time periods. The project uses the frames in GluBam as a complex and multiple systems that are neither hierarchical nor central. This multiplicity and seriality that conforms to a system with repeatable characteristics, makes possible different spatial combinations of the pavilion assembly.

3 THE DESIGN AND CONSTRUCTION OF THE PAVILLON

The design and construction of the pavilion were made by the students of Nanjing Tech University together with some students of the Italian winner team, that went to China in order to assist to the Bamboo Summer School. This team modified the original design idea to simplify the construction. In particular, the size of the pavilion has been reduced to a cube of 2.5 m and most of the elements has been re-designed to reduce the complexity in manufacturing. The original idea was carefully respected in terms of proportions and shape. The final structure was made only of GluBam panels and steel elements for the connection. Each frame was made by two layers of GluBam panel joined by screws. The final structure was obtained by connecting the different frames (a total of 10) using filleted steel bars, bolts and washers. The result is shown in Figure 5. The floor was realized by using two horizontal panels of GluBam. The final structure was easy to be assembled, very strong and can simply be used for exhibition. Moreover, seen the good performances of GluBam, it can also be installed outside.

Figure 5. Final version of the bamboo pavilion at Zhejiang University - University of Illinois at Urbana Champaign Institute, Haining, China: Sketch (left) and photo (center, right).

Figure 6. The closing ceremony at Zhejiang University - University of Illinois at Urbana Champaign Institute, Haining, China. Figure on the left, from left to right: student of ZJU, Dr. Cristoforo Demartino, Dr. Zhi Li, Prof. Erping Li, Prof. Roberto Pagani, Prof. Yan Xiao. Mr. Gabriele Candela, Mr. Stefano Vitale, Ms. Claudia Giorno, student of ZJU and Mr. Ma Ke.

4 THE CLOSING CEREMONY

The Closing ceremony was held on June 30th, 2018 at the Zhejiang University - University of Illinois at Urbana Champaign Institute (see Figure 6). The closing ceremony was chaired by Prof. Xiao Yan, with the important presence of Prof. Roberto Pagani – Science and Technology Counsellor of the Consulate General of Italy in Shanghai – and Prof. Erping Li (dean of The Zhejiang University - University of Illinois at Urbana Champaign Institute). During the discussion, the future of bamboo structure and the best practice to join architecture and engineering were the main topics.

5 CONCLUSIONS AND PERSPECTIVES

This paper summarized the experience of the Sino-Italian "Bamboo #Pavillon Contest_A contest for a Pavilion in Bamboo Technology". The final outcomes of the contest are: i) the creation of an international network between architects and engineers, ii) a great international experience for students characterized by high multidisciplinary, iii) the promotion of bamboo as a construction material and iv) the realization of a prototype of an expositive pavilion realized with bamboo technology. This experience was appreciated by everybody, especially from the student side. In the future, it's important to continue with this experience by enlarging the number of partners involving a large number of Universities and government agencies to better promote the use of bamboo.

ACKNOWLEDGMENTS

Prof. Roberto Pagani, Science and Technology Counsellor of the Consulate General of Italy in Shanghai, is acknowledged for his great support to the contest. The Zhejiang University - University of Illinois at Urbana Champaign Institute is acknowledged for the support provided in the construction of the pavilion. Ph.D. candidate Gabriele Candela (*Mediterranea* University of Reggio Calabria) and Mr. Yue Wu and Ma Ke (Nanjing Tech University) are also acknowledged for helping during the design and construction phases.

REFERENCES

Li, Z., Xiao, Y., Wang, R., & Monti, G. (2014). Studies of nail connectors used in wood frame shear walls with ply-bamboo sheathing panels. Journal of Materials in Civil Engineering, 27(7), 04014216.

Piatkowska, K.K. (2013). Expo Pavilions As Expression Of National Aspirations. Architecture As Political Symbol. Archhist' 13 Architecture Politics Art Conference, Politics in the history of architecture as cause & consequence, Vol. 48, 2013, pp. 20–29.

Tornatora, M., (2017), Learning from pavilion: 100+ 100. Gangemi Editore.

De Capua A., (2015), Towards global architecture. The project between technique and technology, n.14(2)/2015, in Budownictwo i Architektura.

De Capua A., Ciulla V. (2017), Osservatorio P.A.R.C.O. Caratterizzazioni per la qualità ambientale indoor, TECHNE, vol. 1, p. 209–217.

Wang, J.S., Demartino, C., Xiao, Y., & Li, Y.Y. (2018). Thermal insulation performance of bamboo-and wood-based shear walls in light-frame buildings. Energy and Buildings, 168, 167–179.

Xiao, Y., & Ma, J. (2012). Fire simulation test and analysis of laminated bamboo frame building. Construction and building materials, 34, 257–266.

Xiao, Y., Li, Z., & Wang, R. (2014). Lateral loading behaviors of lightweight wood-frame shear walls with ply-bamboo sheathing panels. Journal of Structural Engineering, 141(3), B4014004.

Xiao, Y., Zhou, Q., & Shan, B. (2009). Design and construction of modern bamboo bridges. Journal of Bridge Engineering, 15(5), 533–541.

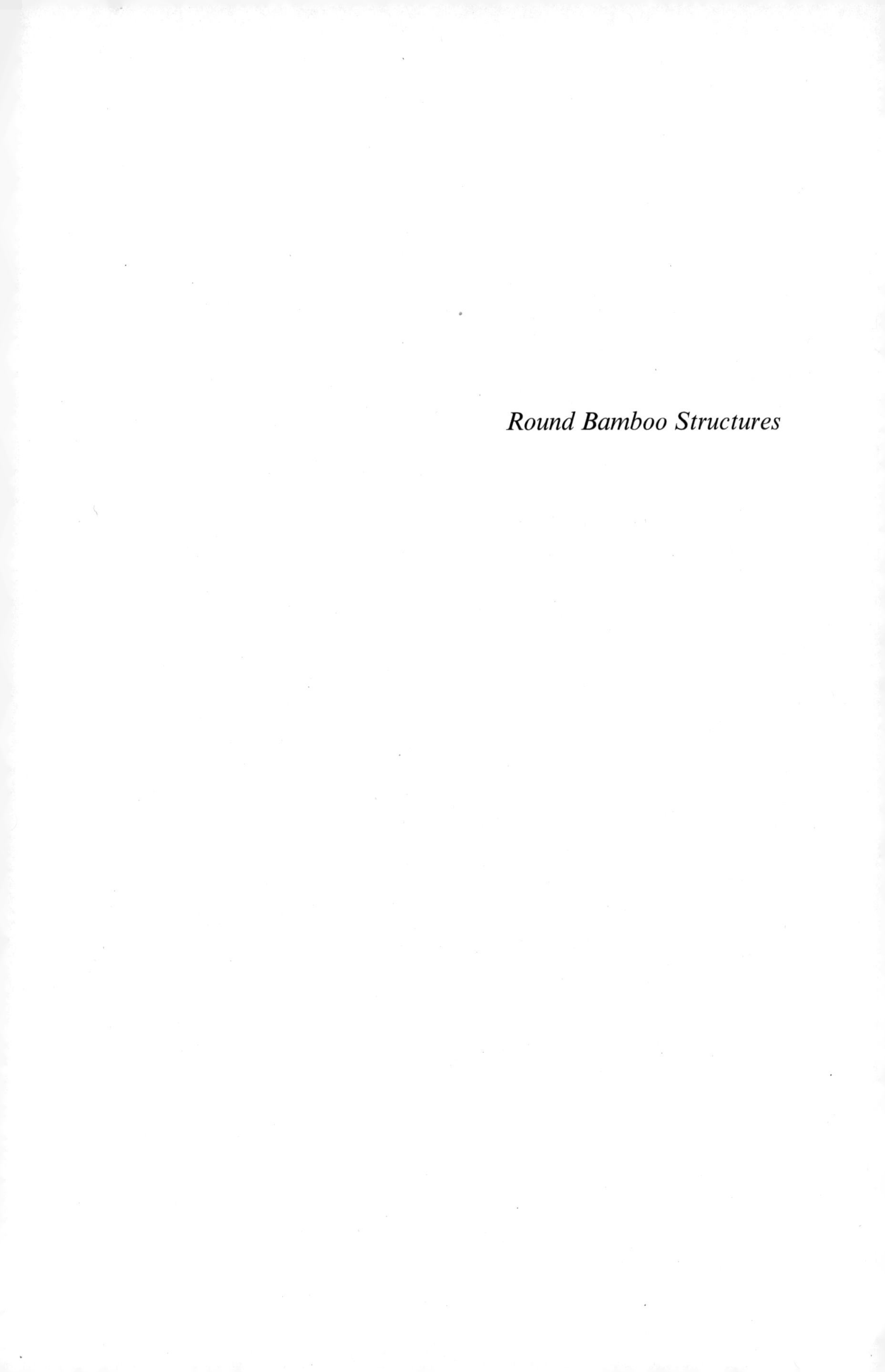

Round Bamboo Structures

Modern Engineered Bamboo Structures – Xiao, Li & Liu (eds)
© 2020 Taylor & Francis Group, London, ISBN 978-1-138-35185-1

Behaviour of *Guadua angustifolia* Kunth bamboo specimens with fish mouth cut under compression parallel to the fibre

C. Takeuchi, E. Ayala & C. Cardona
Universidad Nacional de Colombia, Bogotá, Colombia

ABSTRACT: This paper presents a study of the behaviour of bamboo specimens with fish mouth cut under compression parallel to the fibre. In this research, an experimental stage and several numerical models using the finite element method were performed. In the first stage, specimens with straight cuts at both ends and specimens with one of the ends with fish mouth cut were studied. In both cases, specimens with two types of confinement rings and without any were tested. It was found that in average the resistance to compression parallel to the fibre decreases by more than 30% when having a fish mouth cut. And inclusive, by more than 60% when the specimens did not have confinement rings. However, it was observed that, with the confinement ring, the element ductility improves. Regarding computational analysis, the orthotropic material model presented similar results to the tested samples.

1 INTRODUCTION

In different countries the bamboo has been used for many years in construction as main material. Currently, its design is included in different Standards as the ISO 22156 Bamboo Structural Design (ISO, 2004), the title G Estructuras de madera y estructuras de guadua of the Colombian Standard, NSR 10 Reglamento Colombiano de Construccion Sismo Resistente (AIS, 2010), the Peruvian Standard, Norma Tecnica E100 Bambu (Ministerio de Vivenda Contruccion y Saneamiento, 2012) and the Ecuadorian Stantard, NEC Norma Ecuatoriana de la Construccion Estructuras de Guadua (GaK) (MIDUVI, 2016).

The design of compression elements is included in those Standards, and it is based on the allowable stress calculated with the characteristic value of the compressive strength parallel to fibres, and some safety factors. The fish mouth cut is contemplated in the NSR 10, E100 and NEC Standards as one of the ways to connect elements.

There are some studies related with the compressive strength parallel to fibres of the Guadua angustifolia Kunth bamboo (Takeuchi & Gonzales, 2007), (Correal & Arbelaez, 2010). (Takeuchi, Duarte, Capera, & Erazo, 2012), (Luna, Lozano, Takeuchi, & Gutierrez, 2012), (Takeuchi, Duarte, & Erazo, 2013) and (Luna, Lozano, & Takeuchi, 2014) where different variables have been evaluated; as species, moisture content, source, location at the culm (stem), node (diaphragm) presence, age, among others.

Additionally, to get more knowledge about the behaviour of compressive elements in connections, different researches have been done (Jaramillo & Sanclemente, 2003). Nonetheless, in those experimental studies the elements have failed due to perpendicular to fibre tension or due to the shear parallel to fibre, without identifying the fish mouth cut effects on the mechanical behaviour of the elements.

In this research, an experimental stage and several numerical models, were performed in order to study the incidence of the fish mouth cut on the resistance and behaviour of the compressive elements. In the experimental stage, 50 specimens were tested as follows: 10 with flat faces and without confinement ring; 5 with flat faces and metal clamp; 5 with flat faces and strengthening hoop; 10 with fish mouth cut and without strengthening hoop; 10 with fish mouth cut and metal clamp; and finally, 10 with fish mouth cut and metal strengthening hoop.

As a complement, three numerical models were developed to understand the behaviour observed in the experimental stage, analyzing the stress distribution along the test piece and comparing it with failure modes. The first one represents the flat faces case and the other two involve the fish mouth cut case modelled as isotropic and orthotropic material.

2 EXPERIMENTAL STAGE

2.1 *Specimens manufacture*

Guadua angustifolia culms of 8 cm diameter were selected from Calarca, Quindio in Colombia. 50 specimens were elaborated from five different culms. 20 of them were made with flat faces and the remaining ones with fish mouth cut.

2.2 *Strengthening hoop*

For this stage, two kinds of confinement ring were selected: metal clamp and strengthening hoop. Additionally, it is important to highlight the confinement ring location as seen in Figure 1. In this position, both materials, Guadua bamboo and metal, can work together since the beginning and not when the bamboo has already showed fissures.

For an appropriate monitoring and tabulation of results, the nomenclature mentioned in Table 1 was assigned. In order to compare the results of the different test types, for each culm it was obtain one or two specimens of each type mentioned in Table 1.

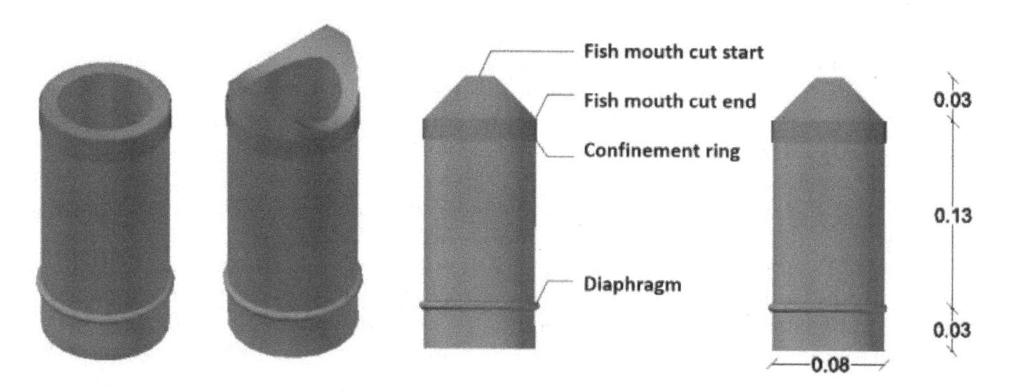

Figure 1. Specimen measures.

Table 1. Experimental stage nomenclature.

Test type	ID	Quantity
Control without confinement ring	CS	10
Control with metal clamp	CA	5
Control with strengthening hoop	CZ	5
Fish mouth without confinement ring	BS	10
Fish mouth with metal clamp	BA	10
Fish mouth with strengthening hoop	BZ	10
TOTAL		50

2.3 *Compression test*

For the test set up, it was necessary to use a semi-circular metallic mould with the same diameter of the specimen, which allowed the load distribution along the fish mouth. On the other hand, the specimens with flat faces did not require any special element, so, a conventional assembly was performed

3 RESULTS

According to machine lectures, stress-Δ/L curves were elaborated, where Δ is the displacement between plates and L is the original length of the sample. Stress-Δ/L curves of specimens without confinements are shown in the Figure 2, with metal clamp in the Figure 3 and with strengthening hoop in the Figure 4.

Additionally, the average of strength of each test type is showed in Table 2.

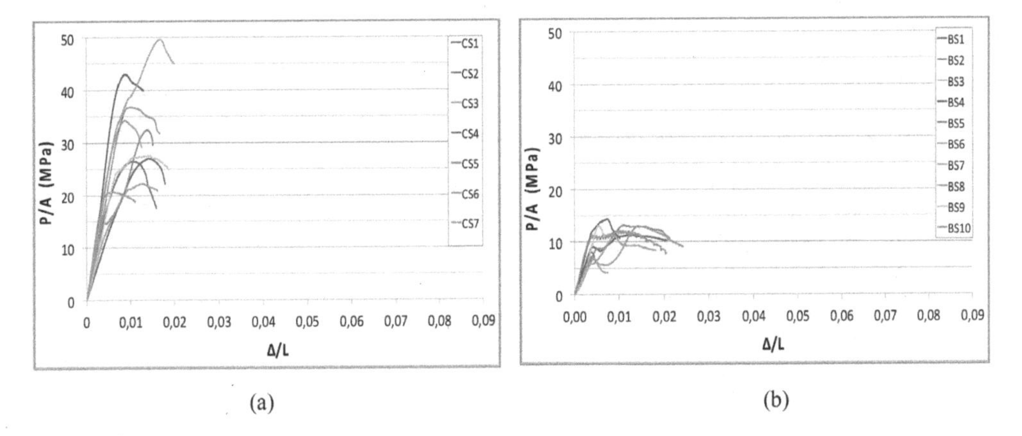

Figure 2. Stress vs Δ/L – Specimens without confinement: (a) Control, CS, (b) fish mouth cut, BS.

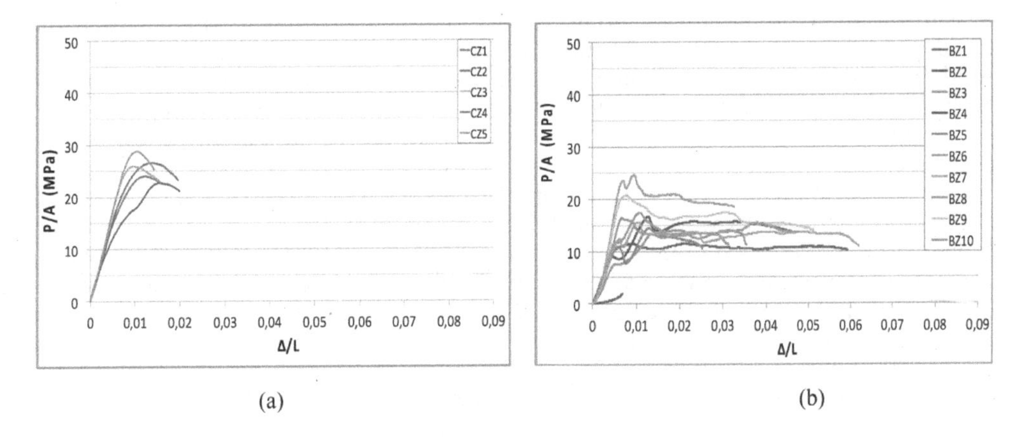

Figure 3. Stress vs Δ/L – Specimens without confinement: (a) Control, CS, (b) fish mouth cut, BS.

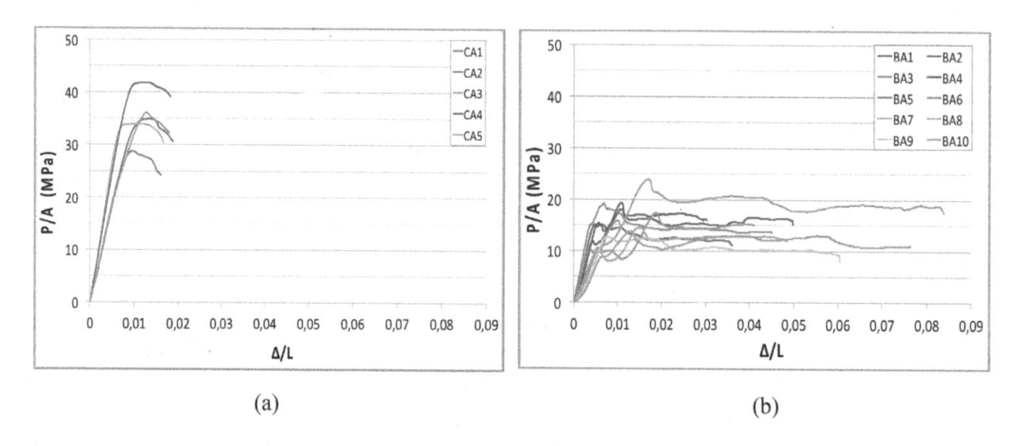

Figure 4. Stress vs Δ/L –Specimens with metal clamp (a) Control, CA, (b) fish mouth cut, BA.

Table 2. Element results.

Test type	ID	Strength average (MPa)
Control without confinement ring	CS	31.97
Control with metal clamp	CZ	25.60
Control with strengthening hoop	CA	35,07
Fish mouth without confinement ring	BS	12.03
Fish mouth with metal clamp	BZ	16.74*
Fish mouth with strengthening hoop	BA	16.94

* Excluding BZ2

4 NUMERICAL MODEL

Elastic analysis for three numerical models with finite element method were developed and allowed the simulation of the behaviour observed in the experimental stage, showing the stress distribution along the test specimen. The first model represents the flat faces case considering an orthotropic material, the second model simulates the specimens with fish mouth cut and uses an isotropic material; the last one, uses the fish mouth cut too, but involving an orthotropic material. The results mentioned correspond to an applied load of 840N, which corresponds to the average stress presented in the flat faces test without confinement.

All models were elaborated in the structural analysis program SAP2000 v14.00, employing "shell" elements to model walls and diaphragms presented in the test pieces. Additionally, "frame" elements were used to transmit the load in the first model. For every model, one direction supports were considered to allow the movement of the specimen in the base. Geometrically, all models were performed considering 8 cm of external diameter and 1.2 cm of wall thickness. Regarding the high, flat faces and fish mouth cut specimens measured 16 and 19 cm, respectively.

As it was mentioned before, for the specimens with fish mouth cut, the load was transmitted using a semi-circular metallic mould with the same diameter of the specimen which allowed the load distribution along the fish mouth. In the numerical model, it was also necessary to include the mould, so that the load was applied over it and then transmitted to Guadua. In Figure 5 the models mentioned before are presented.

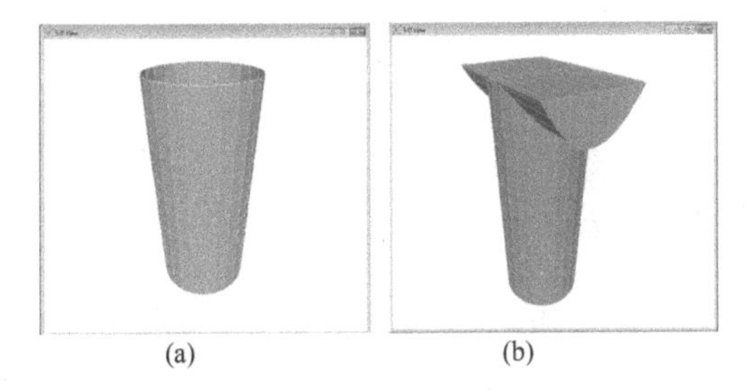

Figure 5. (a) Flat faces test specimen – Model 1 and (b) test specimen with fish mouth cut – Models 2 & 3.

Material properties, used in the analysis, were taken from (Luna et al., 2012) and are presented below.

Isotropic material:
Longitudinal elastic modulus = 11000 MPa
Poisson's ratio = 0.35
Volumetric density = 8 KN/m3

Orthotropic material:
Longitudinal elastic modulus = 11000 MPa
Circumferential elastic modulus = 830 MPa
Poisson's ratio = 0.35
Volumetric density = 8 KN/m3.

5 NUMERICAL RESULTS

In the first model, despite a uniform axial stress distribution along the sample is observed, a little variation is presented in the external face diaphragm zone, due to the presence of the diaphragm that impedes the displacement in this zone.

Regarding the internal face, a little stress increment is presented in the middle zone of the sample. The average compression value is 32.0 MPa. In Figure 6, the stress distribution found in the external and internal faces of the test specimen is presented.

Circumferential stress was also analysed. It was found that tension stresses are presented along the whole specimen, both for external and internal faces. Those are much smaller than compression stresses, having maximum values of 0.2 MPa, considering the internal and external faces, as can be appreciated in Figure 7.

In relation to the second model, which has fish mouth cut and isotropic material, a stress concentration in the lower zone of the fish mouth cut can be observed, reaching a maximum compression value of 56 MPa. Also, a minimum compression value of 10 MPa was located in

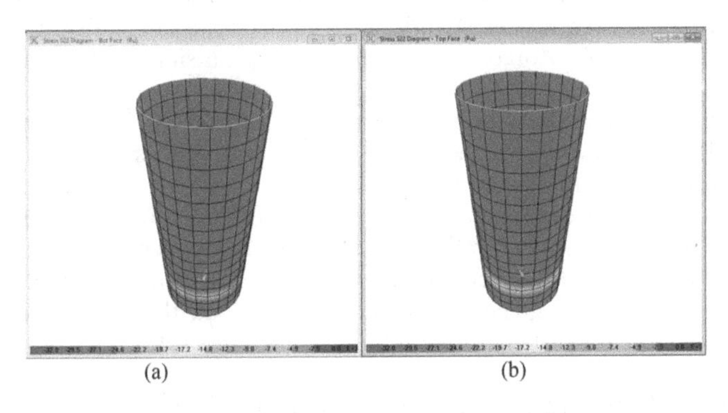

Figure 6. Model 1-Longitudinal stress: (a) external face, (b) internal face.

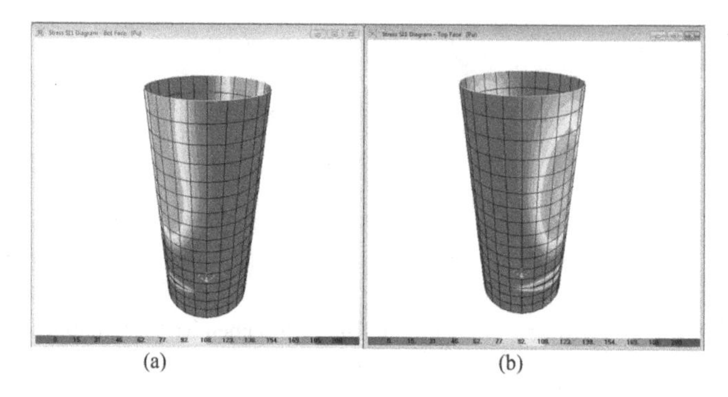

Figure 7. Model 1- Circumferential stress: (a) external face, (b) internal face.

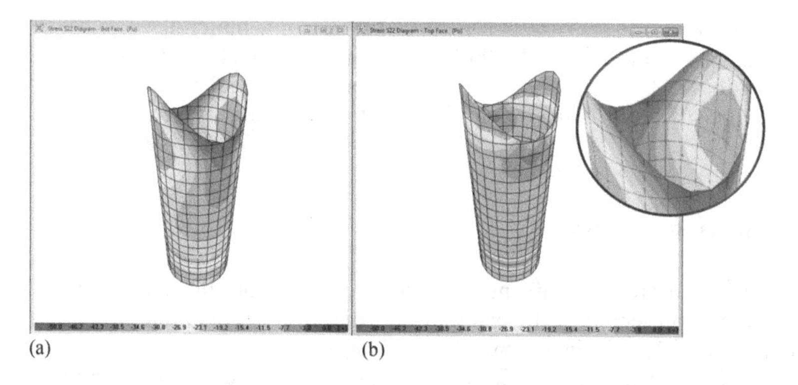

Figure 8. Model 2-Longitudinal stress: (a) external face, (b) internal face.

the upper zone of the mouth (Figure 8). The average stress in the zones outside the stress concentration is 32 MPa (like model 1).

In this case, tension circumferential stress was of 4,8 MPa in the external face and 10 MPa in the internal face, being higher than the tension stress presented in model 1. This stress distribution is presented in Figure 9.

Model 3 differs from model 2 exclusively in the material. In this case, an orthotropic material is considered because it is closer to the Guadua behaviour. As happens in model 2, stress concentration is presented in the lower part of fish mouth, reaching a maximum compression value of 44 MPa (lower than model 2) and, in the upper zone of the fish mouth, a minimum value of 15 MPa (Figure 10). Moreover, the average stress outside the concentration zone continues in 32 MPa.

The tension circumferential stress had a similar distribution than in model 2, reaching values of 0.36 MPa for the external face and 1.1 MPa on the internal face. In Figure 11, the circumferential stress diagram is presented.

In models 2 and 3, compression stress concentration is shown in the lowest part of the fish mouth (Figures 8, 10) and also circumferential tension stress concentration in the middle high of the fish mouth cut. In contrast, model 1 presents a uniform compression stress. According to the results observed, in test specimens with fish mouth cut, the fissure starts in the part of the fish mouth where the highest circumferential tension stress is found, and then it propagates until the diaphragm.

The compression stresses presented in model 2 are higher than in model 3, due to the material used in each model. In model 2, an isotropic material was used, where the elastic modulus

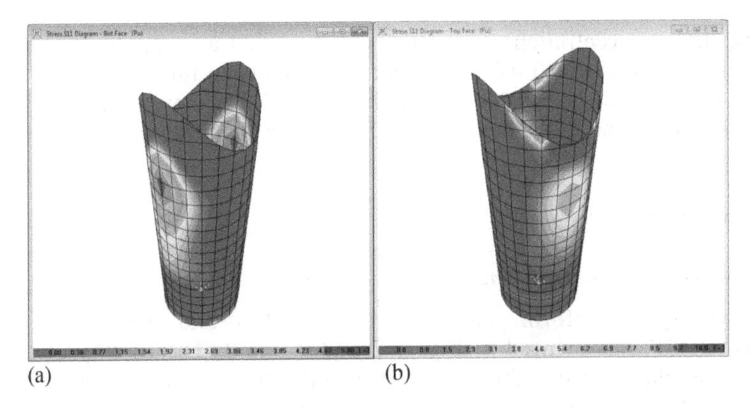

(a) (b)

Figure 9. Model 2- Circumferential stress: (a) external face, (b) internal face.

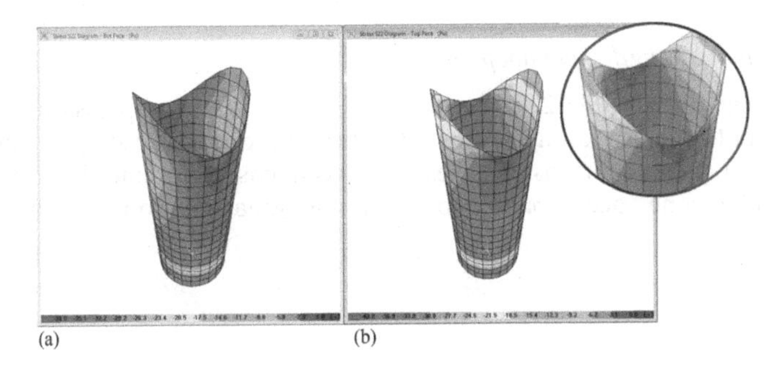

(a) (b)

Figure 10. Model 3- Longitudinal stress: (a) external face, (b) internal face.

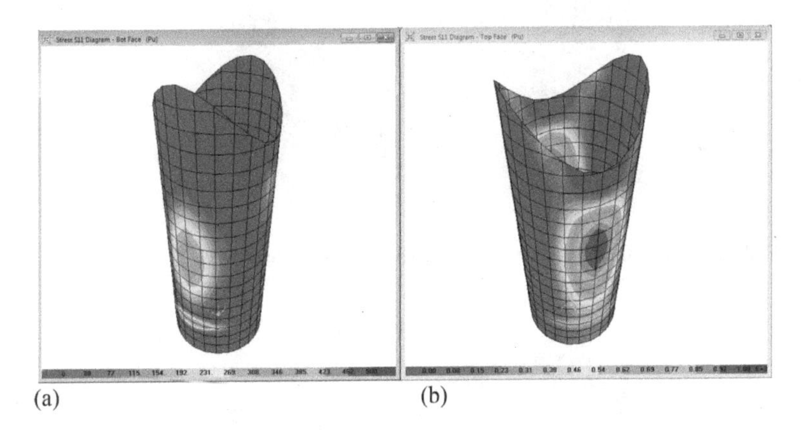

(a) (b)

Figure 11. Model 3- Circumferential stress: (a) external face, (b) internal face.

is the same in all the directions; whereas, in model 3, an orthotropic material was used, where the circumferential modulus is smaller than longitudinal, so there is less radial displacement restriction and less compression stress.

Outside the stress concentration zone, it was observed that average stress is the same in the three models. In the models with fish mouth cut, there is a concentration zone, where the stress is remarkably different in the models. In the third model, the maximum stress is 37% higher than in the first model, and in the model 2 it is 75% higher than the average stress in model 1.

6 RESULTS ANALYSIS

6.1 *Control without confinement ring (CS)*

This group was quite uniform. The deviation was low and the average compression strength was 32 MPa, as evinced in Figure 2 and Table 2. The rigidity of the specimens was extremely similar, but ductility was significantly low due to the lack of confinement ring. Regarding failure type, it was observed, in certain test pieces, a minor squashing on top face, accompanied by tears along the sample until the diaphragm, as can be appreciated in Figure 12.

6.2 *Control without strengthening hoop (CZ)*

As can be appreciated in Table 2, this group had an average compression strength of 25.6 MPa. With respect to kind of failure, it was observed that, due to the strengthening hoop location, tears were not present but instead these the specimens, experienced a kind of strangulation in the middle zone, accompanied by a minor squashing on top face as shown in Figure 13.

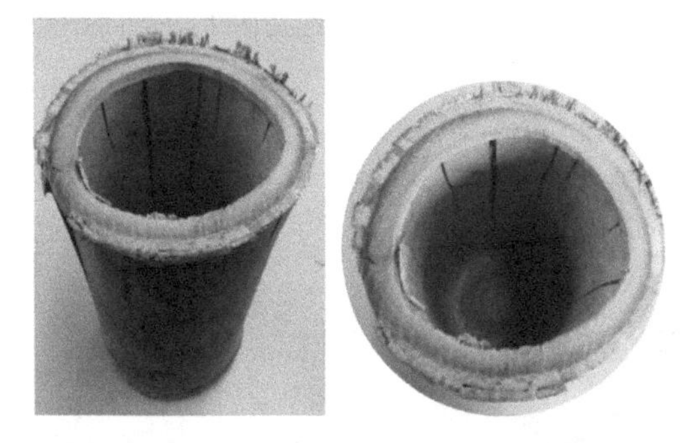

Figure 12. Failure mode control without confinement ring sample (CS).

Figure 13. Failure mode control with strengthening hoop (CZ).

Figure 14. Failure mode control with metal clamp (CA).

6.3 *Control with metal clamp (CA)*

In this group, the average compression strength was 35 MPa higher than the average of the CS tests. Nonetheless, it cannot be affirmed that confinement increases compression resistance, since CZ samples presented lower resistance than control CS specimens. In relation to the failure type, a similar failure to those in control samples with strengthening hoop was presented just below the strengthening hoop, which means strangulation in the middle and dilation at the top, as is presented in Figure 14.

6.4 *Fish mouth specimens without confinement ring (BS)*

According with the finite elements models, the failure was presented in the part of the fish mouth with highest circumferential tension stress (Figure 15).

Regarding the curves behaviour presented in Figure 2, a diminution in the resistance of parallel to fibre compression is evident, reaching an average compression strength of 12 MPa, that is lower than the 32 MPa registered in flat faces samples, nearly the third part.

6.5 *Fish mouth with strengthening hoop (BZ)*

Comparing BZ and control specimens with confinement ring (CA and CZ samples), it is possible to identify a diminution of resistance. However, it is important to highlight that sample BZ-8 reaches a maximum compression value near to 25 MPa, not too far of the maximum value registered in flat faces.

With respect to the failure, the strengthening hoop manages to delay the failure presence, but when the failure is presented, the squashing in the bottom part of the fish mouth is

Figure 15. Failure mode fish mouth without strengthening hoop (BS).

Figure 16. Failure of fish mouth specimens with metal strengthening hoop (BZ).

Figure 17. Failure fish mouth with metal strengthening hoop (BA).

accompanied by some fissures, which, in some cases, can overtake the strengthening hoop or, in other cases, can be visible in the internal face of the specimen (Figure 16).

6.6 *Fish mouth with metal clamp (BA)*

Fish mouth with metal clamp specimens, just like BZ test pieces, evinced a clear resistance diminution compared with CA specimens, but, in this case, it reaches an average compression strength near to 17 MPa versus 35 MPa reached in CA specimens.

In Figure 4, a double increment of resistance due to the strengthening hoop presence can be observed. In relation to failure mode, it can be appreciated something similar to the BZ samples. In addition, it is important to mention that in some cases the strengthening hoop is pushed inside the Guadua bamboo and in other cases it just slides, as happens in strengthening hoop specimens. (Figure 17).

7 CONCLUSIONS

- The compression parallel to the fibre strength of the Guadua is reduced when it has fish mouth cut. The reduction could be more than 40% independently having or no confinement. Considering this, the safety factors for the design with allowable stresses, is reduce drastically.
- The use of confinement ring in specimens with fish mouth cut allows a higher deformation, therefore, a ductile behaviour.
- The existence of tension circumferential stresses on the fish mouth cut, induces the origin of the fissures in the specimens.
- Congruence is presented in numerical simulations that use an orthotropic material.

REFERENCES

AIS Asociación Colombiana de Ingeniería Sísmica (2010). Reglamento Colombiano de Construcciones Sismo Resistente NSR-10. Bogotá. Colombia.

Correal, J.F., & Arbeláez, J. (2010). Influence of age and height position on Colombian *Guadua angustifolia* bamboo mechanical properties. Maderas Ciencia y Tecnología, 12(2), 105–113.

ISO International Organization for Standardization. (2004). ISO Standard 22156, Bamboo Structural Design.

Jaramillo, D., & Sanclemente, A. (2003). Estudio de uniones en guadua con ángulo de inclinación entre elementos. Universidad Nacional de Colombia.

Luna, P., Lozano, J., & Takeuchi, C. (2014). Experimental determination of characteristics values for *Guadua angustifolia.* Maderas Ciencia y Tecnología, 16(1), 77–92.

Luna, P., Lozano, J., Takeuchi, C. P., & Gutierrez, M. (2012). Experimental Determination of Allowable Stresses for Bamboo *Guadua angustifo*lia Kunth Structures. Key Engineering Materials, 517, 76–80.

MIDUVI Ministerio de Desarrollo Urbano y Vivienda. (2016) NEC. Norma Ecuatoriana de la Construcción. Estructuras de Guadùa (GaK).

Ministerio de Vivienda, Construcción y Saneamiento. (2012) Norma Tècnica E100. Bambù.

Takeuchi, C., Duarte T, M., & Erazo E, W. (2013). Comparative Analysis in *Guadua angustifolia* Kunth Samples Solicited to Compression Parallel to the Fiber. Revista Ingeniería Y Región, 10(1), 117–124.

Takeuchi T, C., Duarte T, M., Capera O, A., & Erazo E, W. (2012). Analysis of Variance for Ultimate Strength in Compression Parallel to the Fiber in *Guadua angustifolia* Kunth samples. Revista Ingeniería Y Región, 9, 53–61.

Takeuchi T, C., & Gonzales, C. (2007). Parallel to the Fibre Compression Resistance of *Guadua angustifolia* Bamboo and Determination of the Elasticity Module. Ingeniería Y Universidad, 11(1), 89–103.

Modern Engineered Bamboo Structures – Xiao, Li & Liu (eds)
 ISBN 978-1-138-35185-1

The mechanical properties of cracked round bamboo reinforced with FRP sheets subjected to axial compression

X.M. Meng, Y.Y. Cao & H. Sun
Department of Civil Engineering, Beijing Forestry University, Beijing, China

P. Feng
Department of Civil Engineering, Tsinghua University, Beijing, China

ABSTRACT: The natural round bamboo is a kind of traditional building material, but now barely used for its bad durability and easiness to crack compared to other bamboo productions after secondary operation. In order to improve the safety and durability of the round bamboo structures, the axial compressive test of the GFRP (glass fiber reinforced polymer) reinforcing cracked bamboo was conducted. The round bamboo specimens in the height of 20 cm were divided into four categories: the first without cracks and reinforcement, the second with cracks but without reinforcement, the third with cracks and full GFRP reinforcement, and the last with cracks but only reinforced in the middle height. The bearing capacity and the failure mode were observed and studied. It was found that the GFRP reinforcement could significantly increase the bearing capacity of the cracked round bamboos, and avoid brittle failures through improving the ductility.

1 INTRODUCTION

With the development of the economy and the progress of society, the sustainable development of building materials has become an important research area of civil engineering (Xiao et al., 2009). However, the current main building materials such as steel, concrete, etc., are non-renewable resources, and the manufacturing process of building materials cause enormous environmental pollution and energy consumption (Hui & Zhang, 2007). Bamboo is a kind of environmentally friendly and ecological material (Xiao et al., 2008). It is a natural biomass material with excellent mechanical properties, light weight, high strength, and good seismic resistance (Yu et al., 2011). In addition, bamboo is renewable, degradable, and rich in reserves (Lugt et al., 2006). It just costs about 3 to 5 years to be used as building material (Albermani et al., 2007).

However, natural round bamboo has the drawback of easy cracking parallel to the grain (Low et al., 2006), when the climate is dry or under compression, which will reduce the load carrying capacity of the bamboo structure . The traditional way of strengthening bamboo structure is using steel hose clamps to reinforce, but its corrosion resistance requires careful consideration.

Fiber Reinforced Polymer (FRP) is a new material with the advantages of light weight, high strength, corrosion resistance and convenient construction (Feng et al., 2012). The first use of FRP in civil engineering was mainly strengthening concrete (Smith & Teng, 2002) and steel (Zhao & Zhang, 2007). Therefore, the use of FRP to strengthen the bamboo structures may not only have significant effects, but also reduce the post-maintenance costs. Huang found that CFRP (Carbon Fiber Reinforced Polymer) is better than steel wire reinforcement in strengthening the compressive strength of bamboo, and CFRP can greatly improve its ductility (Huang, 2013). However, there are few studies on the use of GFRP to strengthen bamboo structures in the literature. Therefore, it is necessary to carry out and propose the technical method of GFRP to strengthen the cracked round bamboo and promote its engineering application.

In order to study the reinforcement effects of GFRP on the cracked bamboo columns, the axial compression tests were carried out. The test results would provide an effective reinforcement method for the practical application of natural round bamboo in the future.

2 EXPERIMENTAL PROGRAM

2.1 *Specimens*

The bamboo was collected from Hubei Province. Twelve specimens were cut from the same natural bamboo pole. Each specimen was 200 mm in height. The surfaces were polished. A scanner was used to measure the cross-section area. The specimens were divided into 4 groups: Group A (uncracked and unreinforced), Group B (cracked but unreinforced), Group C (cracked and single-layer full-wrapping using GFRP) and Group D (double-layer half-wrapping using GFRP). The amount of the GFRP sheets was the same for Group C and Group D.

The specimen was named as x-y-z-o, where x, y, z, and o respectively represented the group of the specimens, the position in the bamboo, the crack condition, and the reinforcement method. The position of specimens was divided into the bottom (1), middle (2) and top (3) of the bamboo. The degree of crack was divided into uncrack (UC), slight crack (SLC), severe crack (SC). The reinforcement method was divided into unreinforced method (U), single-layer GFRP and full-wrapping reinforcement (F), double-layer GFRP and half-wrapping reinforcement (H).

Table 1. The details of specimens.

Specimens	Location	Area (mm^2)	Crack	Reinforcement
A-1-UC-U	Bottom	1967	Uncrack	/
A-2-UC-U	Middle	1855	Uncrack	/
A-3-UC-U	Top	1790	Uncrack	/
B-1-SC-U	Bottom	2987	Severe crack	/
B-2-SLC-U	Middle	2452	Slight crack	/
B-3-SLC-U	Top	1498	Slight crack	/
C-1-SC-F	Bottom	2730	Severe crack	Full
C-2-SLC-F	Middle	2298	Slight crack	Full
C-3-SLC-F	Top	1397	Slight crack	Full
D-1-SC-H	Bottom	2629	Severe crack	Half
D-2-SLC-H	Middle	2064	Slight crack	Half
D-3-SLC-H	Top	1350	Slight crack	Half

2.2 *Experimental setup and instrumentations*

This test used strain gages to measure the strain of the test piece. The strain gauges should be pasted in the middle of the test piece as much as possible, one each in the direction parallel to the bamboo fiber and perpendicular to the fiber direction. A vertical strain gauge (V) and a horizontal strain gauge (H) were a group. Each test peace had four groups of eight strain gauges in all and were attached a group every 90o along the loop.

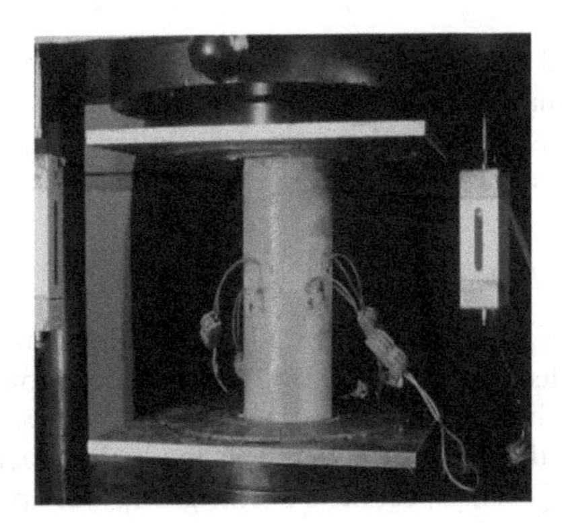

Figure 1. Experimental setup and instrumentations.

3 EXPERIMENTAL OBSERVATION

3.1 *Group A: Uncracked and unreinforced*

At the initial stage of loading, the specimens were slowly compressed, and a slight sound was heard. After a period of time, the uncrack specimens suddenly appeared cracks, and then the cracks sharply ran through the bamboo tube longitudinally, accompanied with a loud sound. When continuing to load, the central part bulged outward and eventually damaged, with bearing capacity slowly decreasing. The group of specimens suffered an overall splitting failure along the fiber direction (Figure 2 (a)).

3.2 *Group B: Cracked but unreinforced*

The experimental observation of Group B was similar to that of the Group A. The difference lied in the original cracks for specimens of Group B continued to develop once starting loading. The cracks also might occur in the originally uncracked part and ran through the bamboo tube along longitudinal direction. The specimens of this group were also the overall splitting failure (Figure 2 (b)), and the damage showed brittleness.

3.3 *Group C: Cracked and single-layer GFRP full-wrapping*

At the initial stage of loading, the specimens were slowly compressed with little deformation. When loaded to the peak load, the local instability occurred in the specimens, and the bamboo wall bulged outward in one direction and inward in the other direction as shown in

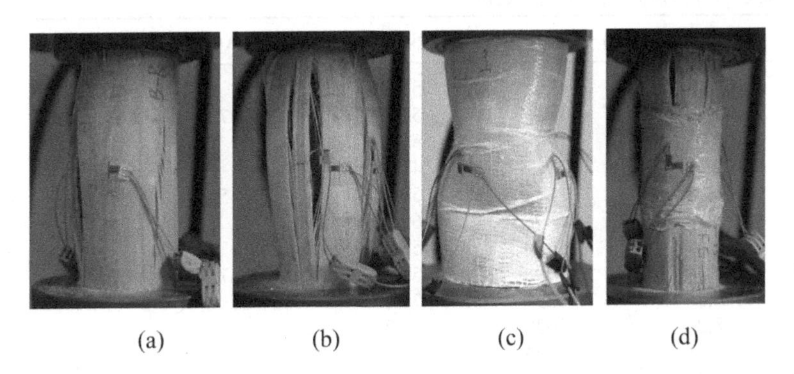

(a) (b) (c) (d)

Figure 2. Failure modes of different groups: (a) Group A, (b) Group B, (c) Group C and (d) Group D.

Figure 2 (c). The GFRP did not fracture along the fiber direction, but due to the structural loss of stability, local deformation occurred and the fiber cracked horizontally. The specimens did not undergo splitting outward with the strong confining effect of GFRP. The specimens of this group showed better ductility.

3.4 *Group D: Cracked and double-layer GFRP half-wrapping*

Similar to the Group C, the specimens of Group D didn't show large deformation at the initial stage. However, when loaded to the maximum load, the double-layer GFRP sheet in the middle height didn't crack, but the original cracks splitted outward at the upper end of the specimens with a loud sound. When continuing to load, the cracked upper part of bamboo bulged outward (Figure 2 (d)). The specimens of this group were partially damaged.

4 RESULTS AND DISCUSSION

4.1 *Load-displacement relationship*

According to the experimental data, the results of the axial compressive strength of specimens are calculated and shown in the Table 2 below. The typical load-displacement relationship curves are shown in the Figure 3 (a).

It was found in Table 2 that the peak load of the specimens decreased from the root to the top of the bamboo pole. However, the compressive strength parallel to the grain direction was increased from the root to the top of the bamboo pole.

Compared with the specimens of Group B, the compressive strength of the specimens of Group A was relatively larger than that of Group B, and the curves suddenly decreased. It indicated that the original crack was an unfavorable factor affecting the compressive strength of the round bamboo. The primary cracks in the bamboo joints would develop together with the new cracks, resulting in a sharp drop in the bearing capacity. Furtherly, the compressive strength of the specimens of Group C and the specimens of Group D were improved, and the improvement of Group C was greater than that of Group D. The results showed that the use of GFRP to strengthen the round bamboo could effectively improve the bearing capacities. Moreover, the full-wrapping was better than the half-wrapping when strengthening the bamboo structure.

If the compressive strength of Group B was defined as the control group, it could be seen that the compressive strength of specimens reinforced with GFRP almost reached the compressive strength of the uncracked specimens. For the full-wrapping specimens, the failure mode showed ductile deformation, but the failure of the half-wrapping was less ductile. Therefore, the full-wrapping method was more usefull.

Table 2. The experimental results of specimens.

Specimens	Peak load (kN)	Area (mm^2)	Strength (MPa)	Ave. strength (MPa)	Increase(%)
A-1-UC-U	109.89	1967	55.87		
A-2-UC-U	107.24	1855	57.81	57.20	25%
A-3-UC-U	103.68	1790	57.92		
B-1-SC-U	122.97	2987	41.11		
B-2-SLC-U	123.74	2452	50.46	45.76	0
B-3-SLC-U	90.73	1498	60.56		
C-1-SC-F	152.67	2730	55.92		
C-2-SLC-F	137.24	2298	59.72	57.82	26%
C-3-SLC-F	106.54	1397	76.26		
D-1-SC-H	128.22	2629	48.77		
D-2-SLC-H	115.58	2064	56.49	55.69	22%
D-3-SLC-H	83.46	1350	61.8		

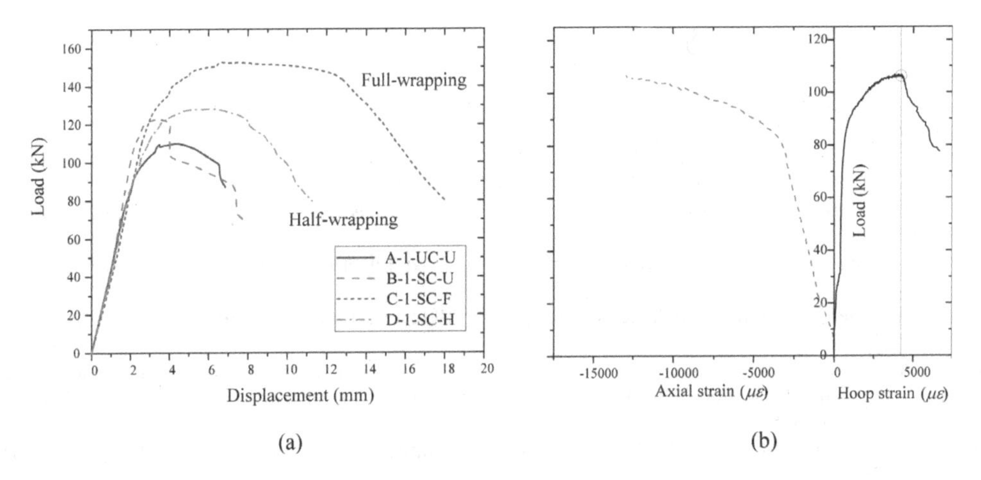

Figure 3. The experimental results of the specimens: (a) Load-displacement relationship; (b) load-strain relationship of specimen C-1-SC-F.

4.2 *Load-strain relationship*

Since the single-layer GFRP full-wrapping was used to confine the specimens, the hoop strain was related to confinement effect. The specimens no longer bulged outward, but the local instability occurred. The strain gauge was attached to the GFRP surface. The strain values might not correctly reflect the deformation, but it actually measured the strain of the GFRP in the initial stage.

The load-strain relationship of specimen C-1-SC-U was shown in Figure 3(b). It could be seen clearly that the peak hoop strain was less than 5000 $\mu\varepsilon$, which was much less than the ultimate strain for GFRP material. That is to say, the GFRP used in the specimens wasn't fully used and the confinement effect was relatively weak. Therefore, the reinforcement method using GFRP sheets could be better design.

5 CONCLUSION

In this paper, the axial compression tests were carried out to study the reinforcement effects of GFRP on the cracked bamboo columns. The conclusion is as follows:

(1) The wrapping method changed the failure modes from outward split to inward deformation, and improved the peak loads and markedly improved compression deformation.
(2) The confinement effect of GFRP sheets wasn't fully used. The optimization of GFRP reinforcement might be an important issue.
(3) The reinforced cracked round bamboos showed better mechanical behavior, which might promote the development and application bamboos in constructions and buildings.

ACKNOWLEDGMENTS

The authors acknowledge funding supported by the Fundamental Research Funds for the Central Universities of China (No. BLX201706), and supported by Major Science and Technology Program for Water Pollution Control and Treatment (No. 2017ZX07102-001), and supported by the National Natural Science Foundation of China (No. 51278276 and No. 51522807).

REFERENCES

Albermani, F., Goh, G.Y. & Chan, S.L. 2007. Lightweight bamboo double layer grid system. *Engineering Structures*, 29, 1499–1506.

Feng, P., Bekey, S., Zhang, Y.-H., Ye, L.-P. & Bai, Y. 2012. Experimental study on buckling resistance technique of steel members strengthened using FRP. 12, 153–178.

Huang, G.Q. 2013. *The Experimental Research and Analysis on The Mechanical Properties of Reinforced Bamboo*. Shanghai Jiaotong University.

Hui, L.I. & Zhang, Y.K. 2007. Analysis on Ecological Architectural Material of Bamboo. *Building Science*.

Low, I.M., Che, Z.Y., Latella, B.A. & Sim, K.S. 2006. Mechanical and Fracture Properties of Bamboo. 312, 15–20.

Lugt, P.V.D., Dobbelsteen, A.a.J.F.V.D. & Janssen, J.J.A. 2006. An environmental, economic and practical assessment of bamboo as a building material for supporting structures. *Construction & Building Materials*, 20, 648–656.

Smith, S.T. & Teng, J.G. 2002. FRP-strengthened RC beams. I: review of debonding strength models. 24, 385–395.

Xiao, Y., Inoue, M. & Paudel, S.K. 2008. Modern bamboo structures.

Xiao, Y., Shan, B., Chen, G., Zhou, Q., She, L. & Yang, R. Developing Modern Bamboo Structures for Sustainable Construction. Iabse Symposium Bangkok Sustainable Infrastructure Environment Friendly, 2009.

Yu, D., Tan, H. & Ruan, Y. 2011. A future bamboo-structure residential building prototype in China: Life cycle assessment of energy use and carbon emission. *Energy & Buildings*, 43, 2638-2646.

Zhao, X.L. & Zhang, L. 2007. State-of-the-art review on FRP strengthened steel structures. 29, 1808–1823.

Modern Engineered Bamboo Structures – Xiao, Li & Liu (eds)
© 2020 Taylor & Francis Group, London, ISBN 978-1-138-35185-1

Ethiopian vernacular bamboo architecture and its potentials for adaptation in modern urban housing: A case study

A.D. Dalbiso & D.A. Nuramo
Ethiopian Institute of Architecture Building Construction and City Development (EiABC), Addis Ababa University, Addis Ababa, Ethiopia

ABSTRACT: Ethiopia is not only endowed with a huge bamboo resource but also has a very rich traditional bamboo housing construction techniques, which have been in use arguably for more than thousand years. These construction practices have several benefits including provision of affordable houses for millions living in rural areas of the country, employment creation, reducing burden on natural forests for housing construction, and income generation. Two types of bamboo species are existing in Ethiopia namely, *Oxytentra abyssinica* commonly known as lowland bamboo and *Yushania alpine* commonly known as highland bamboo. Both species are used for construction of residential houses where several vernacular architectural design approaches are exhibited including the Sidama house in Southern Ethiopia. This study particularly focuses on exploring how the existing bamboo construction techniques in Sidama region, provide a platform for provision of houses for emerging towns in bamboo growing areas in Ethiopia. Data collection including observation and personal interview was conducted in the region. The data was processed using both qualitative and quantitative methods where construction techniques and the overall process was analyzed technically and areas of improvement were identified. Based on the findings a modified building constriction system and construction process is proposed.

1 INTRODUCTION

Vernacular architecture is a term used to categorize method of construction using traditional knowledge and ingenious locally available resources to address local needs. This type of architecture provides highly responsive techniques towards addressing climatic constraints and show high amount of adaptability and flexibility (Shanthi Priya, Sundarraja, Radhakrishnan, & Vijayalakshmi, 2012). Vernacular traditions lead a way towards the sustainable built environment. The valuable lessons from vernacular can be integrated with the modern to produce sustainable designs. Vernacular traditions can also be used as a design tool for housing programs however the designing of these settlements need understanding users' way of life, social and cultural values (Shikha & Brishbhanlali, 2014).

In Africa, Ethiopia, Kenya and Uganda possess most of the bamboo resources, according to the world bamboo resources assessment report (Zhao et al., 2018). Among the three countries, 86% of the African bamboo resource is distributed in Ethiopia. Two indigenous species of bamboo in East Africa are *Yushania alpina* (highland bamboo) and *Oxytenanthera abyssinica* (lowland bamboo).

The reliance on masonry and reinforced concrete building systems has made even the most modest housing units too expensive for the majority of urban dwellers. In contrast; rural housing is more affordable due to the common application of indigenous construction material like bamboo. This is especially true in tropical parts of Ethiopia where bamboo can be harvested from the wild in large quantities. The potential adaptation of indigenes and abundant building material like bamboo as an alternative construction material for modern structures could alleviate the cost of housing units in urban Ethiopia.

Ethiopia's population is increasing at a very high rate. The latest population and housing census conducted in 2007 by the Central Statistics Agency shows the country's population at 74 million; 50.5% male and 49.5% female (CSA, 2007). United Nations population projections show this number passing the 100 million mark by 2015 and reaching close to 190 million by 2050 (United Nations Department of Economic and Social Affairs, 2017).

Although the least urbanized nation in Africa, in the past ten years Ethiopia's urban population has been growing rapidly. This growth is mainly attributed to the influx of rural dwellers in search of a better life and employment opportunities in urban center. Currently 17% of the country's total population lives in urban centers, and in a typical growth pattern of least urbanized nations, its urban population is growing at 5% per year making Ethiopian cities some of the fastest growing in the world (Nyamongo, 2017).

2 METHODOLOGY

Case studies focused on the vernacular bamboo architecture of the Sidama People in southern part of Ethiopia and its adaptability to modern urban housing is investigated. Both primary and secondary data are collected through field observations, literature review and interviews of the local artisans. The steps which were followed during the field study include information gathering, on-site observation, interview with local people, photography and visualization of some building elements by means of sketches.

The findings from the study are grouped into seven sub-categories. The sub-categories include: sustainable design culture; sustainable land use; designing for durability; construction materials and techniques; construction waste management; energy efficiency; and indoor air quality; and

2.1 *The Sidama People*

The Sidama people are found in the Southern Nations Nationalities and Peoples Region (SNNPR) of Ethiopia whose capital city is Hawassa. The district has an area of 7672 km^2 and a population of 2,954,136 (CSA, 2007). It is located between 5°45′ and 6°45′N latitude and 38° and 39°E longitude. The altitude varies from 1500 to 3500 m asl. The areas above 2000 m asl are generally suitable for growing bamboo. The people in this region have access to natural forest highland bamboo, but majority of them traditionally practice private bamboo farming.

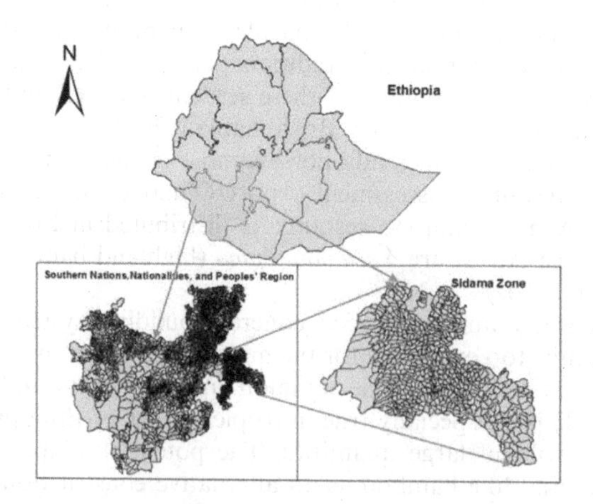

Figure 1. Map of the study area (Sidama Zone).

2.2 *House form, orientation and basic construction systems*

2.2.1 *Spatial layout of Sidama house*

The Sidama house is a unique beehive shaped structure that is finished by fixing a layer of undifferentiated woven bamboo onto the structure. Partitions are also made of woven bamboo. Most of the houses have two entrances, a back and front entrance. The back entrance is meant for use by the cattle and sheep while the front entrance by the people.

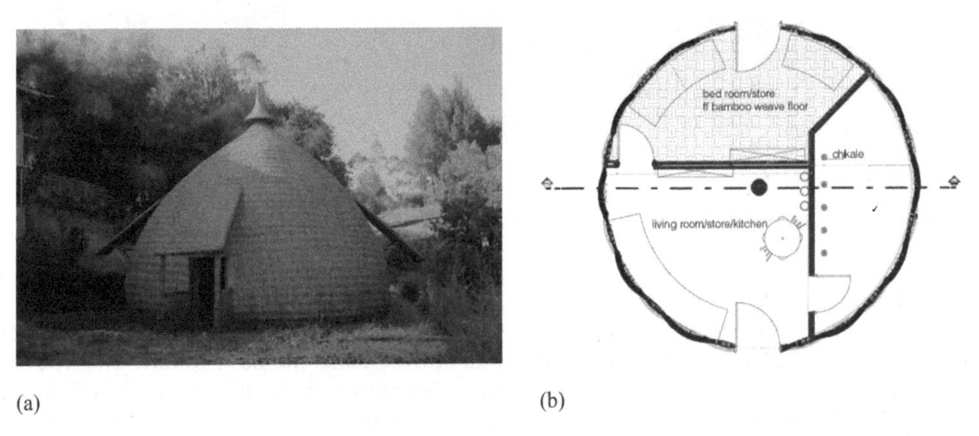

Figure 2. Sidama House: (a) View; (b) spatial layout.

Traditionally the house has 3 parts; the residential, the cooking space and the 'Arkata' for a cattle and crop store. The interior is accessed through a porch and it's divided into 'Olico' which is further divided into 'Holge' (parents sleeping area) and 'Bosalo' (sleeping area for children's and guests). The 'Bosalo' is also used for storing production materials and other small items. The central pole 'Helicho' is very important in Sidama culture. The older people 'Chamesa' wouldn't enter the house unless the house has central pole.

2.2.2 *Foundation and structure*

Juniper tree poles are used as a component for the foundation. It is applied for the foundation because of its longevity and being proven to protect against pest attack like termites. The poles are imbedded to the ground '*Mokolicho*' is used as reinforcement bars. Thin strands of bamboo are placed around the perimeter of the house. It can be placed up to 100 to 150 cm underground. '*Hicho*' are smaller bamboo that are not split and are used for weaving the interior and exterior walls.

Figure 3. Juniper tree poles used for foundation.

(a)

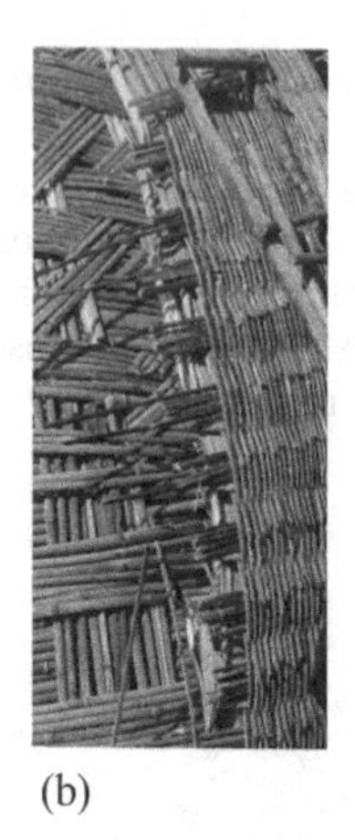

(b)

Figure 4. Wall section: (a) Interior skin of wall; (b) exterior skin of wall.

2.2.3 *Wall*

The construction of the walls is made by splitting the bamboo to smaller strips and uses whole bamboo with smaller diameters. The interior wall of the Sidama hut has two different types of patterns. ('Hilo' and 'Himbiro') The 'Hilo' pattern follows a linear weaved form, which has a basic parallel appearance. Whereas the 'Himbiro' pattern has a diagonal weaved pattern.

The interior wall of the hut has a wall finish known as 'Lemicho'. The wall finish can have an extra layer called 'Chicha'. The 'Chicha' can be applied in different colors, patterns and sizes. The material used for waterproofing the hut 'Honche' or bamboo sheath (Figure 4). It is harvested from the skin that the bamboo shoot sheds. It is placed in four layers all around the hut. The exterior wall of the Sidama hut is weaved with a pattern called 'Fuko'; which has an overlaying effect on the façade.

2.2.4 *Roof*

A beautiful dome or dome like shape is made for the roof with a triple layer of bamboo splits and ropes for structural support, culm sheaths for insulation and rain protection. A woven basket like cover on the outside ties everything together and protects the house against wind (Figure 5).

2.3 *Methods of bamboo treatment*

Smoking of house: dry grasses, paddy stems, dry leaves, dry bamboo leaves, rice husk and some amount of green leaves and branches are used for smoking of house. The smoking is done under the observation for 2 - 3 days.

3 PROBLEMS AND ISSUES WITH TRADITIONAL BAMBOO HOUSES

- Most of the present dwellers of bamboo houses belong to the socially and economically weaker sections in the society. Probably due to their poor economic condition. They opt for low cost bamboo houses and hence are of low quality. Thus economic conditions of the dwellers must be improved to construct quality bamboo houses.
- Demand of bamboo for the construction sector is more than the total supply of bamboo. Quality bamboos, at low cost must be made available to people who wish to construct bamboo houses.
- Most of the bamboo community is aware of the construction technology and traditional bamboo preservation techniques. If financial help is provided they can build reasonably safe buildings for themselves with little training on latest technology interventions.

(a) (b) (c) (d)

Figure 5. Dome structure: (a) Weaving of bamboo strips to form dome shape; (b) erection of the dome; (c) construction of door shade; (d)false umbrella cover for aesthetics.

- Bamboo may be grown in wastelands and mid and lowland areas and homesteads should be encouraged. All this can be included in the plantation programs of the Forest Departments in rural and tribal areas.
- Rationing of bamboos in areas where its supply is less or price is more can be thought of. More bamboo depots may be established by the Forest Departments in different parts of the country so that the dwellers can buy bamboo easily.
- The government should include bamboo as a construction material in their housing scheme thereby upgrading the status of bamboo houses. Economical technology for preservative treatment of bamboo should be popularized for construction. Some model bamboo houses suitable for local conditions and weather can be constructed to popularize the same among the people.

Table 1. Lessons from vernacular architecture of sidama people.

Aspects	Parameters	Particulars
Socio-Cultural	Family Structure	Nuclear family structure
	Fairs & Festivals	The festivals are governed by nature, like the festival of 'Fitche Chambalala'
	Community Participation	A group of men from the 'Chinacho' which are a crew of skilled weavers, are called on for the construction of the homes with the help of the other members of the community.

(*Continued*)

Table 1. (*Continued*)

Aspects	Parameters	Particulars
	Belief and rituals	The vernacular buildings reflect the environmental and climatic circumstances, which are well integrated into religious or spiritual convictions, and strongly tied to the ancestors and the social community, ritual and symbolism
Ecological	Building with nature	The form and the structure of the Sidama house is mainly the result of the topography, climate, culture and material used.
	Family Name/Identity	Family names and identities are associated with nature. Therefore they never harm them
Architectural	Site selection	The houses are built on plateau surrounded by the hills, site is usually close to the source of water. The construction is done on non-fertile land.
	Climate responsive	The house form is evolved as per the climatic condition. Less openings are provided due to extreme temperature. The bamboo material gives an optimum indoor thermal qualities in all seasons
	Settlement pattern	It has circular settlement pattern with the community space at the center
	Materials	Almost entirely built from bamboo
	Construction Method	A beautiful dome or dome like shape is made for the roof with a triple layer of bamboo splits and ropes for structural support, culm sheaths for insulation and rain protection. A woven basket like cover on the outside ties everything together and protects the house against wind.
	Aesthetics	The interior of the walls and doors are decorated with bamboo matts
Economic	Livelihood	Contribution to the household income from bamboo harvesting
	Resource management	Judicial use of materials
	Waste management	Building materials provided from the closest environment do not generate waste while in use or after use because they are organic in nature

4 CONCLUSION

The form and the structure of the Sidama house is mainly the result of the material used. The onion shape and circular plan are created through the flexibility of the bamboo plant. They prefer to do most of the construction work in groups. A group of men from the 'Chinacho' which are a crew of skilled weavers, are called on for the construction of the homes with the help of the other members of the community. The people in the Chinancho select a leader called 'Murcha' to supervise the whole construction process which takes 2-8 weeks. Members who do not cooperate are excluded from the group.

REFERENCES

CSA. (2007). The 2007 Population and Housing Census of Ethiopia: Statistical Report for Addis Ababa City Administration. Mortality. https://doi.org/10.1016/j.wasman.2015.02.013.

Nyamongo, I. (2017). STATE OF AFRICA'S POPULATION 2017 Keeping Rights of Girls, Adolescents and, (March).

Shanthi Priya, R., Sundarraja, M.C., Radhakrishnan, S., & Vijayalakshmi, L. (2012). Solar passive techniques in the vernacular buildings of coastal regions in Nagapattinam, TamilNadu-India - A qualitative and quantitative analysis. *Energy and Buildings*. https://doi.org/10.1016/j.enbuild.2011.09.033.

United Nations Department of Economic and Social Affairs. (2017). World Population Prospects: The 2017 Revision - Data Booklet. *Population Division*. https://doi.org/10.1017/CBO9781107415324.004.

Zhao, Y., Feng, D., Jayaraman, D., Belay, D., Sebrala, H., Ngugi, J., ... Gong, P. (2018). Bamboo mapping of Ethiopia, Kenya and Uganda for the year 2016 using multi-temporal Landsat imagery. *International Journal of Applied Earth Observation and Geoinformation*. https://doi.org/10.1016/j.jag.2017.11.008.

Modern Engineered Bamboo Structures – Xiao, Li & Liu (eds)
© 2020 Taylor & Francis Group, London, ISBN 978-1-138-35185-1

Innovations in round bamboo construction

W. Cai
Zhujing Bamboo Company, Huzhou, Zhejiang, China

T. Li
Nanjing Tech University, Nanjing, Jiangsu, China

ABSTRACT: At present, existing round bamboo constructions have following limitations: first, the service life of raw bamboo is short, only about 2-3 years, and the corresponding treatment method is difficult to meet the international environmental protection requirements. Secondly, the construction technology is backward, the connection and support methods are simple and weak, cannot meet the requirements of the modern structural design. Finally, traditional round bamboo building products have insufficient innovation in design and low added value. With the purpose of solving the above problems, the bamboo-based product design company of the author focused on the research and development of round bamboo construction in the following aspects: 1) anti-corrosion and anti-mildew treatment of raw bamboo, 2) connection system, and applied for the related patents. At the same time, this research also introduces the actual engineering design cases based on it.

1 INTRODUCTION

Round bamboo construction has a long history in countries with abundant bamboo resources, such as China, India, and South America. Bamboo, as an environmentally friendly construction material, has advantages such as fast-growing, high strength, low density, low production cost, ease of manufacturing etc., The round bamboo building is graceful, flexible and diverse, which is not only a place for people to rest, but also a unique landscape line of the cultural tourism scenic spot, with beautiful appreciation value and the value of garden culture and art. However, the development of round bamboo construction is yet to be well developed to meet modern construction requirements, due to (1) mildew and insects, (2) difficulty in connection and support, (3) insufficient design innovations. Therefore, this paper discusses the methods of trying to solve the above problems in the round bamboo construction industry and some projects are introduced.

2 ADVANTAGES OF ROUND BAMBOO CONSTRUCTION

According to the author's design practice, there are five obvious advantages to develop round bamboo construction in China: 1) rich raw resources: bamboo is one of the forest resources. There are about 107 species and over 1300 species of bamboo in the world, mainly distributed in tropical and subtropical regions. The bamboo forest in China covers an area of over 6 million hectares, among which the most valuable bamboo is moso bamboo. The bamboo forest in China accounts for 90% of the total bamboo forest in the world, and China has extremely rich raw bamboo resources. 2) carbon emission: the round bamboo building conforms to the principle of ecological design. Raw bamboo can be produced within 3 to 6 years, which is economically and environmentally friendly and in line with China's sustainable development strategy, compared with the production cycle of timber of every 50 or 60 years. The raw bamboo building can also greatly reduce the proportion of the structure to the building area and greatly improve the space utilization rate. Raw bamboo is a completely degradable material, which is different from traditional building materials such as steel, concrete, and stone, which will be left for thousands of years without degradation. Compared with bamboo forests and trees in the same area, the carbon dioxide absorption is 4 times as much as the forest, the oxygen release is 1.35 times as much, the negative oxygen ion contribution rate is 1.3 times as much as the forest, and the soil retention capacity is 1.46 times as much as the forest. Therefore, bamboo is the most effective controller of the global greenhouse effect. 3) strength: the unique hollow, ribbed soft inside, together with the hard-outside structure of raw bamboo tube is a great invention of nature. From the roots to the top, the coherent fibers of raw bamboo make them the hardest plants in the world, earning themselves a "steel plant" reputation. The tensile strength of bamboo is about 2 - 25 times that of lumber, and the compression strength 1.5 - 2 times. The tensile and compression strength of a whole bundle of raw bamboo is beyond imagination. 4) visual appearance: the color of bamboo is soft and quietly elegant, the texture is clear and exquisite, giving people the double enjoyment of vision and psychology. The round bamboo architecture has the Oriental culture flavor, with a high degree of discrimination. 5) flexibility: compared with other building materials, the round bamboo building is designed and constructed more flexible and can be maintained regularly by replacing individual damaged parts.

3 LIMITATIONS OF TRADITIONAL BAMBOO CONSTRUCTION

As shown in Figure 1, traditional processing technology cannot effectively protect raw bamboo from mildew and insects, leading to the short service time of raw bamboo structure. The raw bamboo contains abundant organic nutrients: 1.5%- 6% of protein, 2% of soluble sugar, 2.2%- 5.18% of starch, 2.18%- 3.55% of fat and wax, and more than 30% of water. These nutrients pose potential problems of mildew and moth in raw bamboo. Meanwhile, due to the complex density distribution of bamboo fiber bundles, the nutrients distribution is uneven. The nutrient located in the inner circle of the raw bamboo is much higher than that of bamboo outer skin and fiber, rendering the protection and treatment of the raw bamboo much more difficult than that of wood and segmented bamboo.

The traditional domestic raw bamboo treatment method basically employs the cooking method as shown in Figure 2. In this method, 2 kg of fire alkali is added to every cubic meter of water, within which the raw bamboo is completely soaked for about 30 minutes after boiling. This treatment can hardly solve the problems of anti-mildew and anti-insect. Bamboo joints in the raw bamboo make it difficult for the alkali to penetrate vertically. Moreover, the outer green fiber bundles are small, fine and covered by silicon and wax on the surface, making it even harder for the solution to reach the core parts. Finally, the solution contains harmful chemicals that do not meet environmental requirements.

At present, domestic traditional bamboo building mainly uses two kinds of materials: green bamboo and white bamboo treated by "cooking method", whose service time normally does not exceed 3 years. The connection and support methods are simple and weak.

Figure 1. Raw green round bamboo.

Figure 2. Cooking treatment method.

Figure 3. Traditional original bamboo construction.

The raw bamboo is basically a hollow cylinder, the connection and the support point can't simply use mortise and tenon joints as timber structure. Traditional construction technology uses the hemp rope, iron wire to tie up or direct gun nailing, neither solid nor beautiful. Therefore, Traditional bamboo building products have insufficient innovation in design and limited added value.

4 INNOVATIONS IN ROUND BABOO CONSTRUCTION AND CASE-STUDY

Mildew prevention and pest control of the raw bamboo (Patent No.: 201610549284.9) should be the priority problem to be solved for the development of modern round bamboo structures. The service life of treated raw bamboo can be greatly improved: the service life for indoor space has reached more than 30 years, for semi-indoor space (with top without wall) has reached more than 20 years, and for complete outdoor space has exceeded more than 10 years, as shown in Figures 4 & 5. Metal connections, as shown in Figures 6 & 7, is another important aspect to establish a reliable structure system for round bamboo structures. Raw bamboo connecting, supporting and other core technologies, which endows the raw bamboo with a new shape, aesthetic feeling, and reliability. Several round bamboo construction projects with those innovative solutions in raw bamboo treatment and connections are given in Figures 8 to 13.

Figure 4. Raw bamboo decorative structure (Patent No.: 201610549283.4).

Figure 5. Raw bamboo decorative lamp (Patent No.: 201620735495.7).

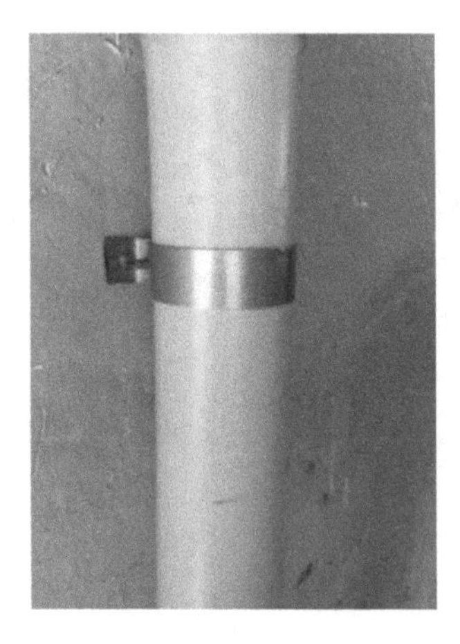

Figure 6. Raw bamboo fastener (Patent No.: 201620735496.1).

Figure 7. Raw bamboo support (Patent No.: 201620735494.2).

Figure 8. Kunqu society in Xibang Village.

Figure 9. Township hall in Shang Village.

Figure 10. Yunnan sea ancient ferry terminal.

Figure 11. Shenyang panda house.

Figure 12. Circular landscape, Anji, Zhejiang.

Figure 13. ZhuLi pavilion, Anji, Zhejiang.

5 CONCLUSIONS

The key to the development of the round bamboo building is the following two aspects: first, to solve the problem of mildew and insect control of the raw bamboo, so that the use life of the raw bamboo can reach the market requirements; secondly, to focus on the connection, support, function development and other aspects of raw bamboo technology to develop unique round-bamboo based modern structural products.

REFERENCES

Bamboo awning township hall in Shang Village 2018. https://www.archdaily.cn/cn/893253/an-hui-shang-cun-zhu-peng-xiang-tang-su-po-jian-zhu-gong-zuo-shi.

Cai, W. 2016. A treatment process of anti-corrosion and anti-mildew for raw bamboo. Invention Patent, No. 201610549284.9, National Intellectual Property Administration, PRC.

Cai, W. 2016. A primary bamboo decoration structure. Invention Patent, No. 201610549283.4, National Intellectual Property Administration, PRC.

Cai, W. 2016. Raw bamboo decorative lamp. Utility Model Patent, No. 201620735495.7, National Intellectual Property Administration, PRC.

Cai, W. 2016. Raw bamboo fastener. Utility Model Patent, No. 201620735496.1, National Intellectual Property Administration, PRC.

Cai, W. 2016. Raw bamboo support. Utility Model Patent, No. 201620735494.2, National Intellectual Property Administration, PRC.

Kunqu opera society in Xibang Village 2017. https://www.archdaily.cn/cn/892681/xi-bang-cun-kun-qu-xue-she-zhong-guo-jian-zhu-she-ji-yan-jiu-yuan-star-ben-tu-she-ji-yan-jiu-zhong-xin.

Shenyang panda house 2017. http://www.sohu.com/a/141030232_167180

Yunnan sea ancient ferry terminal 2017. http://www.sohu.com/a/156148331_668172.

Modern Engineered Bamboo Structures – Xiao, Li & Liu (eds)
© 2020 Taylor & Francis Group, London, ISBN 978-1-138-35185-1

Bamboo culm buildings

R. Mostaed Shirehjini
Nanjing Tech University, Nanjing, Jiangsu, China

Y. Xiao
Zhejiang University-University of Illinois at Urbana Champaign Institute, Jiaxing, Zhejiang, China
Nanjing Tech University, Nanjing, Jiangsu, China
Department of Civil Engineering, University of Southern California, Los Angeles, CA, USA

ABSTRACT: Becoming more aware of environmental impact associated with producing common construction materials (e.g. wood, concrete, steel) is drawing on ancient tradition of using natural organic materials in architecture and structural engineering. Bamboo, as one of the oldest natural building materials, currently has been proved to be a green substitute to wood, concrete and steel due to its exceptional natural properties such as flexibility and higher specific strength and stiffness. However, due to the lack of knowledge and understanding of bamboo's potentials as building material, this versatile material continues to be undervalued by designers and communities. With conducting desk research including case studies, this study investigates the use of bamboo culms in traditional and modern architecture and highlights the design opportunities and challenges. The current research is expected to put forward a deeper understanding of using bamboo culm as a building material and the results to be of interest and benefit architects, structural engineers, and decision makers involved in the construction projects.

1 INTRODUCTION

1.1 *Context*

The growing population and increasing demand for construction activities has given rise to environmental concerns and depletion of natural resources. The previous researches show that production of most of the man-made building materials such as concrete (Flower, 2007) and steel (Olmez, 2016) require a huge amount of virgin materials, water and energy for processing purposes, transportation and construction works which makes them responsible for a significant environmental impact. To bridge the gap between nature and our built environment, it is important to select bio-based rapidly renewable materials which have low emitting potential (Ampofo-Anti, 2010).

Bamboo, has been proved to be among the most environmental friendly in the world (Liese, 1998) (Escamilla, 2014) (Xiao, 2009) with 20 times less environmental impact than currently used building materials (Mahdavi, 2010). Bamboo is strong. According to Jules Janssen, "the weight of a 5,000-kilogram elephant can be supported by a short bamboo stub with a surface area of just 10 square centimeters (Ariyani, 2012)."

However, lack of strong interest in bamboo (Hidalgo Lopez, 2003) and conceiving it as poor man's timber, moreover, insufficient information on bamboo and its potential of use in construction "have been a drawback in its promotion (Zoysa, 1988)."

The current research attempts to put forward a deeper understanding of bamboo culm in its unprocessed form and using it in traditional and modern building design. With conducting desk research and literature case studies on traditional and modern bamboo buildings, this paper highlights the design opportunities and architectural aspects of bamboo culm buildings.

It is expected that the results to be of interest and benefit to architects, structural engineers, and decision makers involved in the construction projects.

1.2 *Bamboo as a plant*

Bamboo is not a tree. It belongs to a grass family with about 1600 species and 121 genera. Bamboos naturally grow in tropical, subtropical and mild temperature regions, and they are mostly found in Asia, Africa, South and Central America (Lobovikov, 2007). Bamboo species and their growth behavior can fall into two distinct groups: running bamboos which can spread horizontally, and clumping bamboos that extend into a short distance around the perimeter (Hidalgo Lopez, 2003).

The bamboo culm is a series of hollow cylindrical internodes and intersections called nodes. The nodes produce roots and branches (Hidalgo Lopez, 2003). The internodes have a culm wall of different thickness, around a hollow space. The culms are denser in outer layer than innermost part, and generally, as they grow, they take a taper shape (Liese, 1998).

Bamboos are the fastest growing plants recorded in the world. In order to achieve the fast growth expectations climate, soil and topography condition are the main factors. Most bamboos grow at temperatures between 9°C and 36°C, on a "well-drained, fertile soil." (Hidalgo Lopez, 2003) They reach to their maximum strength at age three and become mature.

Bamboo culms can be green, yellow, black, red and white in color. In addition, there are some species with strips on their skin (e.g. Guadua angustifolia var. bicolor and Bambusa Vulgaris Vitatta) (Hidalgo Lopez, 2003).

"Most of the bamboo species have round culm section." However, there are also species with natural square or triangular culm sections. Bambusa ventricosa McClure have bottle-shaped culms. "Freak bamboos" are species that undergo deformation during their growing phase (e.g. Turtle shell and Buddha's face Bamboos) (Hidalgo Lopez, 2003).

Environmentally, bamboo plant is very friendly. It grows rapidly and can be harvested more frequently without depletion or soil decay (Lobovikov, 2007). Bamboo regrows automatically after being cut on its existing roots (Hidalgo Lopez, 2003). It "can grow in hilly sloppy lands that are not suitable for agriculture or forestry (INBAR, 2017)." Bamboos release 35 percent more oxygen than equivalent trees and it performed even better than fast growing Chinese fir tree in accumulating carbon. "Their sturdy rhizomes and roots regulate water flows and prevent erosion (INBAR, 2014)." They can significantly reduce the burden on forestry resources and decline in biodiversity, and offer many other environmental benefits (Lou, 2010).

1.3 *Bamboo as a building material*

To make best use of bamboo species it is important to understand their physical and mechanical characteristics and the way they respond as a building material.

"The cellulose fibers in bamboo act as reinforcement similar to reinforcing steel bars in concrete (Ghavami, 2003)." The distribution of fibers increases from inside to the outside. In terms of load support and stiffness against deformation, bamboo is about the same as steel and the second best after steel respectively.

Density, in most types of bamboos lies between 700–800 kg/m^3. The culm density increases from innermost towards the outer part, and from the lowest point to the top. The upper part of the culm is smaller but also the strongest part of the culm with highest bending strength. Bamboo culm shape is "smart" in such a way that the bending stress caused by the wind load, distributes evenly along the height and withstands the wind force (Sutnaun, 2011).

"Bamboo does not have rays like timber" which cause mechanically poor point for timber. However, the cavity in the bamboo culm make this advantage ineffective, and bamboo in most cases possess a weaker shear stress comparing to timber.

The lightweight and hollow form has given bamboo culms a perfect stiffness and functionality during earthquake. "Bamboo houses have survived quite near to the epicenter of a 7.5 Magnitude earthquake." In such houses, even if the roof or walls collapse during earthquakes, it is unlikely to kill or injure the family members. Furthermore, to fix or replace the damaged parts is inexpensive and fast (Almendral, 2014).

Compared to woods, bamboos have less natural durability. An untreated bamboo that is exposed and in contact with soil, can be used maximum for three years. This lifespan can be increased to 10-15 years if the culms are well used and stored in good conditions. Preservation is another solution, which is best to enhance bamboo's durability (Janssen, 2000).

2 CONSTRUCTION WITH BAMBOO

A bamboo building "is a building that uses bamboo as the major material for the main structure and components of the building (Mardjono, 2002)." In such buildings, bamboo is used in two main fashions. One is to use bamboo culm without undergoing significant changes (to use it almost as it is), and another way is using engineered or processed bamboo panels. This paper is a study on building with unprocessed bamboo culm.

Developing an effective bamboo building design with a good performance is possible, when both traditional and modern methods are applied (Mardjono, 2002).

2.1 *History*

Bamboo has given human beings "shelter and protection" throughout the history. In Southeast Asia, bamboo was used for construction by extinct human species one million years ago. Prehistoric "bamboo dwelling or funeral chambers" reconstructed in America dates back 9,500 years ago, and the history of bamboos covered with plaster used in bahareques spans through 3,500 years to the present (Hidalgo Lopez, 2003). At any places where bamboo grows, variety of bamboo houses with indigenous technologies have arisen and evolved across the world from east to west. Stilt bamboo houses so called Nipa hut in Philippines inspired "Father of the American skyscraper" William Le Baron Jenney to build the first skyscraper with lightweight structure, when he saw Nipa huts' "flexibility and lightness" and how they endure and function well during the disasters.

It is estimated that today, over a billion of people worldwide (INBAR, 2015) live in traditional bamboo houses. However, using bamboo as a building material is traditionally considered a characteristic of poverty within people and the communities. For example, in India and Philippines only the lower class use bamboo. In Colombia, many bamboo buildings are disappearing with a dramatic speed, and becoming replaced with concrete and brick constructions (Hidalgo Lopez, 2003).

Instead, in the past two decades, with increasing interest among professional designers, architects and engineers over using unprocessed bamboo culm in modern green architecture, the number of public and private bamboo buildings with different forms and sizes are increasing. These buildings are mostly designed and built in Southeast Asian countries such as Indonesia, Vietnam and Thailand.

Today, bamboo as a building material still needs to fulfill engineering requirements. With this respect, a standard bamboo test method is essential to reliably provide fundamental material properties based on bamboo species, geometry, weathering and treatment methods, and present it as a concrete design guidance to different species (Harries, 2012).

2.2 *Traditional bamboo housing*

Generally, traditional buildings/houses are the structures built based upon the knowledge passed on from one generation to another within the family members or communities, as part of their building tradition in that region. Worldwide – mostly where bamboo grows naturally – depending on the climatic zones and available local bamboo species, different traditional building methods and typologies are developed over the centuries.

According to Rapoport in (1969), "indigenous, or folk, building is the direct and unself-conscious translation into physical form of a culture's needs and values. Included within the indigenous tradition in housing are traditional and vernacular buildings which have been built for centuries with little monetary cost, without imported materials, using existing skills, and with renewable resources (Rapoport, 1969)." For example, in coastal areas where floods are common (e.g. in Bangladesh, Southeast Asian countries), also in the hilly areas (to build on non-flat surfaces), bamboo houses are traditionally lightweight huts raised on stilts. In India, stilt bamboo houses such as Kutchcha house and Garo house, Riang houses are typical traditional hill dwelling (Lala, 2017).

The importance of ventilation in the wet tropical region can be achieved through the use of "split-bamboo-low-ventilating walls sheltered by eaves" and by cross ventilation (Kahn, 1978).

In Latin America, particularly in Peru, Ecuador, and Colombia, where some rural construction has remained the same since before the Spanish conquest, bamboo has been used for centuries in walls, lightweight ceilings, tiles, and matting (Hartkopf, 1985). Guadua bamboo (Guadua angustifolia) has long been the most common material for low-cost housing in Colombia and Ecuador." Three of the most popular methods for walls construction employed by the indigenous builders of Latin America are; "1- bahareque (Figure 1): which is a bamboo frame coated with a mud/straw mixture applied to the surface of a structure. 2- quincha (Figure 2): as a sprung strip construction in which flexible strips of bamboo are woven together to provide a base for the mud/straw plaster which is applied over it 3- Unfinished bamboo poles: used in the lowest cost constructions. Whole culms or flattened culms."

Traditional houses have a relatively short life span, as a rule, but their styles persist and many of them are still being constructed (Van Dine, 1977)." However, these bamboo houses – usually found in rural areas – end up being as fragile temporary shelters, with a very low level of protection for its inhabitants.

Hence, although bamboo is a great material for construction, there are still obstacles in many aspects such as cultural and technical be addressing and solving systematically. The major problems and limitations of traditional bamboo houses, such as durability and stability issues, are resulted from using old methods without the help of modern tools and new technologies. For example, lack of knowledge on bamboo culms treatment before being used, and to maintain bamboo buildings is one of the many reasons causing durability problem. Moreover, using traditional systems for connecting the joints make the structures weak and fragile (Figure 3).

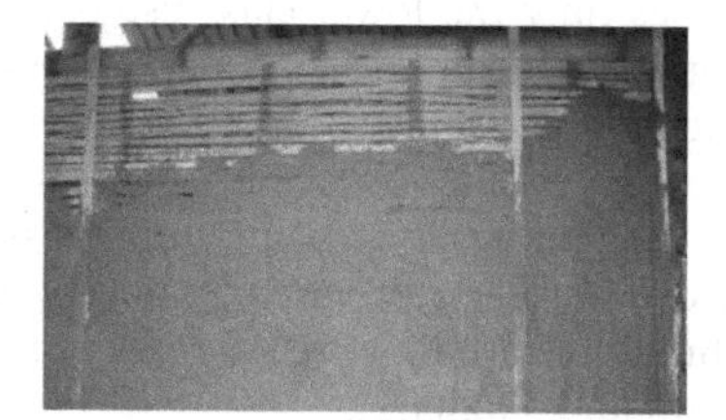

Figures 1–2. Pictures from left to right; bahareque and quincha bamboo wall systems.

Figure 3. Traditional bamboo houses on stilts built over non-flat ground surfaces or water body.

2.3 *Bamboo culm in modern architecture*

Bamboo has the image of being the building material of the poorer class. The low quality houses that people build with bamboo, not to mention the "beliefs they hold about bamboo as a sign of poverty" is leading people in countries with a long track record in using bamboo as a building material (e.g. Colombia, Philippines, India), today prefer concrete rather than bamboo. Besides, since bamboo is not a standardized material to European people, it is inconvenient for them to build with bamboo.

"Nevertheless some famous architects and engineers already made their experiments with this natural product. Simón Vélez is one of the most famous pioneer architects who used bamboo in innovative ways as a main building material. Simón Vélez and his own "well-trained" team constantly draw upon past successes and failures in detailing, and they have developed a model for building experimental structures. The very first time that Simón decided to build with bamboo was when one of his friends asked him to build a big roof bamboo stable for his horses, without knowing much about how to join the bamboo culms, as the nails and screws damage the culms easily. Finally, he decided to "use bolts and secure the nodes", and creating concrete foundation to support the bamboo sticks (Villegas, 2003). He designed lasting joints and a long spans and cantilevers, without using structural calculations and only with drawing sketches in a notebook.

Velez's contributed to modern bamboo architecture in two ways: first, improving the image of bamboo construction among the society (Velez, 2000); and second with introducing new technologies in bolts connections, parallel connections, orthogonal connections and angle joints injecting concrete at the end of the bamboo culms creating long spans and cantilevers (Rodríguez-Camilloni, 2009). The Temporary Cathedral is one of the early attempts in modern bamboo architecture, designed by Simón Vélez in Colombia in 1999 (Figure 4).

The literature about bamboo in modern architecture is hard to find. However, demonstrating the existing projects gives a good chance to convince people bamboo is a great versatile material for design purpose. The increasing improvements in construction with bamboo, has brought more contributions to the diversification of bamboo architecture as to form, shape, texture and line. In some modern buildings, bamboo culms function as claddings and non-load bearing walls and ceilings, creating aesthetic appeal, endless possibilities in form and so on (Figure 5).

Figure 4. Temporary Cathedral at Pereira, Colombia. By Simon Velez.

Figure 5. Bamboo culms used for shades and claddings.

Figure 6. Bamboo buildings put striking architectural ideas on display, and demonstrate the application of bamboo culms as key element in architectural and structural design of the contemporary architecture.

In bamboo culms buildings, culms are the main material used for load bearing and non-load bearing components such as columns, beams, trusses, walls, roofs. Bamboo culms can be used straight and shape rectangular or Geodesic forms, or they can be circular and curved and create organic and complex structural/architectural forms. Construction with bamboo with the help of new methods and technologies allows large spans, fluidity, and lightness and gives resist earthquakes (up to 9.0 magnitude) feature to the building. Bamboo displays beautiful structures and creates sustainable and luxurious spaces with positive environment (Figure 6).

3 CASE STUDIES ON BAMBOO CULM BUILDINGS

Considering the fact that there is a lack of literature indicating the possibilities and challenges of using bamboo in modern architecture, to illustrate the existing modern bamboo buildings and their analysis can be of help. Some leading modern bamboo buildings are as follows;

3.1 *Ecological children activity and education center at Six Senses, Soneva Kiri resort – Thailand*

The case study on "Ecological children activity and education center at Six Senses– Thailand" is extracted from (Henrikson, R. & Greenberg, D., 2011) and ("24H", 2009)

The project inspired by nature, designed and built by 24H Architecture aiming to "change the bad reputation of bamboo and inspire people and architects" by displaying the bamboo's stunning performance as a building material in modern architecture. The main material used in this project, is bamboo. The interior made from local cultivated River Red Gum wood and rattan structural elements for the inner domes. 24H Architecture designed and built it in the shape of a floating giant manta ray amongst the trees, with a big belly!

During the design process, in addition to the 3D computer model, a sizable 1: 30 scale physical model built and tested in wind-tunnel by engineering company Ove Arup.

For the main structure used Pai Tong bamboo (after being tested on tension, compression, shear and bending) in lengths up to 9 m and a diameter of 10–13 cm. The secondary roof and 'belly' structure is made from Pai Liang bamboo (Bambusa multiplex) in 4 m lengths and a diameter around 5 cm. In this structure, there is a combination of modern and traditional joining technologies used. For the main load-bearing bamboo structures they have used bolted joints, and filled bamboos with cement injections to prevent bamboo from splitting. For the secondary

Figure 7. Manta-ray inspired bamboo building in Thailand.

roof- and belly-structure, bamboo dowels were mainly used in combination with rattan which was winded around the bamboo bundles. The roof consists of a ceiling of split bamboo with a waterproof membrane on top, covered with bamboo shingles. Adopting all bioclimatic aspects for design, the 8 m roof cantilevers acting like a big umbrella providing shade and protection from the heavy rains. The open design with the translucent elevated rooftop and setback floors allow a natural airflow and daylight inside, limiting the building's energy consumption."

(Figure 7).

The structure is 12 m in height and 28 m in width. A system for joining culms was invented to make them long enough, using dowels lashed with ropes. For the interior they used natural finishes such as river red gum and rattan, cloth with resin-adhered soil, or textures derived from bamboo sawdust.

3.2 *Green School, Bali – Indonesia*

Brigitte Shim had conducted this on site case study in May 2010 (Shim, 2010).

3.2.1 *Concept, design, construction*

"In 2007, John and Cynthia Hardy's decided to start a new school shaping both its educational curriculum and its built form." The Green School buildings, is a sustainable campus designed creatively and built in Bali, with full respect to traditional vernacular architecture of the region creating a modern place for invention and experimentation with bamboo which is a locally available material.

Bali has a humid tropical climate with two distinct monsoonal wet and dry seasons, and little temperature variation throughout the year (average 26–30°C). Bamboo species are plentiful in Bali along with many other native and cultivated species. The ambitious project "Green School" designed by multi-disciplinary integrated design team, using trial and error approach, research, hard work, various skill levels "deep commitment". Architects, graduate architects, jewelry designers, sculptors, structural engineers, bamboo experts, master builders worked together in a variety of ways. Nevertheless, to create such giant complex and irregular buildings with low-tech, using local artisans and craftsmen, and their ingenuity was the key.

Total floor area (ground floor and upper floors) of the project is 7,542 square meters. Exposed bamboo structures are "expressed and celebrated". Extensive roofs are everywhere with alang alang (a local material) cover to protect the space in both wet and hot seasons. Very few walls and windows are the key for inviting natural ventilation inside (Figure 10).

Figure 8. Alang alang roofing and full bamboo culm structure has created a positive environment.

Figure 9. A Cluster of structural bamboo columns and bamboo arches in Green School.

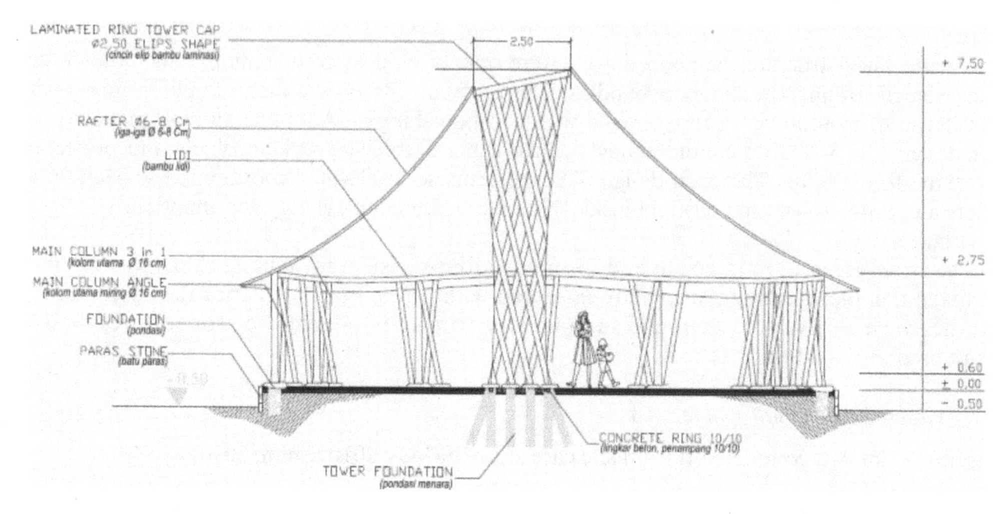

Figure 10. Kindergarten section in Green School.

A cluster of vertical bamboo columns creates long span arches. Giant dendrocalamus asper bamboo, adopted as the primary structure forming three interlocking trusses. The load of the structure transfers to the foundation. "The primary structure is anchored to the foundations by an innovative method of lacing river rocks and bamboo with reinforced steel connected to the concrete foundations." In order to have the buildings withstand wind forces, the connection between bamboo culms to the ground tapped and filled with cement. The lighter weight bamboo species, like Gigantochloa Apus used for the secondary structure and rafter elements fixed to the primary bamboo Petung with bamboo pins. Using overlapped alang alang gives added strength to the structural components. To raise a three-story building, a group of bamboos with 16–18 meters high as structural members shaped the heart of the building (Figure 9).

A1: 1 scale model for testing the joints, and a 1: 100 physical construction model for whole construction provided with details. Then, the model transformed into a computer model to test and analyze different loadings such as axial and earthquake response. The 18 m column free span and 14 m in height multipurpose facility of Gymnasium, constructed with four structural bamboo arches each consisting three Petung bamboo.

The main buildings in the Green School campus are free from any wall or door. To provide interior enclosures, they have used bamboo infill panels, natural latex saturated canvas and single glazing. For buildings rendering and finishes, designers applied the traditional system of roofing in Indonesia, using tough resilient grass named alang alang for thatching.

All materials incorporated in the Green School are locally available 99–100% natural, and as well recycled. They used bamboo throughout the Green School, and mixed local mud with 15% cement to form all the floors in the campus. Moreover, volcanic stone used for its permeable nature for paving all of the pedestrian pathways and parking areas.

The main structural columns are from local bamboo species with their particular length. For upper floors of buildings, they used 100% bamboo splits connected together with bamboo pins, using no glue or chemical finishes. Only for those structurally critical points, they have used bolts to join the bamboo and injected the hollows with cement.

3.3 *Technical assessment*

The openness of bamboo building of the Green School has allowed ample of natural and diffused daylight to enter, and provides cross ventilation and air movement. This helps cooling down the building, humidity dry out easily and preventing moisture to build up.

The large roofs with generous overhangs, keep the rain away and protects the building. However, those big umbrellas are vulnerable to high wind uplift and the structural systems have to provide an "appropriate anchorage and resistance."

Lightweight bamboo culms used for columns, long span bamboo arches, has made the Green School buildings well responsive to when earthquakes occur. The Green School campus is located on steep, hillside slopes. It creates a natural protecting area from any flood. "The Ayung River also provides a readily available and accessible water supply in case of fire."

For all structural bamboo used at Green School, they have done preservation with leaving the culms drawn in the Borax bath for 4–6 weeks. Again, they sprayed the bamboo culms with an environmentally friendly waterproof coating.

The Green School buildings monitoring is an important part of the project and its maintenance. A team of leading structural engineers through a proactive approach conducts a visual examination for checking the joints throughout the campus. Based on building performance they make the modifications if required and "Selective replacement is anticipated." Alang alang roofing with 6–10 years replacement cycle and bamboo based on "conservative estimates" has 20 years life cycle.

4 DISCUSSIONS AND CONCLUSION

The existing literature claims that bamboo plant is a boon for the environment and besides, its structural capacity is comparable to steel, concrete and timber. However, using it is mostly based on traditional knowledge, and hence leading to low quality housing, has undervalued the bamboo's potentials and contributions in construction. On the other hand, there is insufficient updated literature on bamboo in modern architecture, as well lack of design guidance and standardization for bamboo structures.

Nevertheless, there are a number of modern bamboo buildings constructed, which expose the abundant potentials this amazing green material possess. To that end, the considerations derived from the studies cases include but not limited to; right bamboo selection and treatment methods, a climate responsive design, trial-error approach and research-based analysis, using physical models (usually sizable) for test and providing 3D computer models for structural analysis, bringing together multidisciplinary experts, craftsmen and artisans, and continues maintenance and monitoring. Bamboo can be used straight, curved or in organic forms and new connecting joint methods allow bamboo to adopt a variety of structural/architectural massing. Bamboo columns can hold multistory buildings and bamboo arches can create large spans. Bamboo is from nature and give us a unique opportunity to build in harmony with the nature, and create sustainable, luxurious building with welcoming colors and positive environment.

REFERENCES

Almendral, A. 2014, June 30. Typhoons Are Getting So Strong That Focus Is Shifting Away From Recovery. Retrieved Oct 22, 2018, from Next City: https://nextcity.org/daily/entry/typhoons-are-getting-so-strong-that-focus-is-shifting-away-from-recovery.

Ampofo-Anti, N. 2010. Material selection and embodied energy. Green building handbook South Africa: the essential guide, 3.

Ariyani, K. 2012, June 10. Bali goes green with bamboo buildings. Retrieved November 1, 2018, from Science X: https://phys.org/news/2012-06-bali-green-bamboo.html.

Escamilla, E. 2014. Environmental impacts of bamboo-based construction materials representing global production diversity. *Journal of Cleaner Production*, 117–127.
Flower, D. 2007. Greenhouse gas emissions due to concrete manufacture. *The International Journal of Life Cycle Assessment*, 282–288.
Garcia-Saenz, M. 2012. Social and cultural aspects of constructions with bamboo. *Tenth LACCEI Latin American and Caribbean Conference*, (pp. 1–7). Panama City.
Ghavami, K.A. 2003. Bamboo: functionally graded composite material. *Asian Journal of Civil Engineering* (Building & Housing), 1–10.
Ham, A.C. 1990. Bamboo housing in Costa Rica: an analysis of a pilot program (Doctoral dissertation). Adrienne C. Ham.
Harries, K.A. 2012. Structural use of full culm bamboo: the path to standardization. International Journal of Architecture, *Engineering and Construction*, 66–75.
Hartkopf, V. 1985. Tecnicas de construccion autoctonas del Peru. Washington, DC: Agency for International Development.
Hebatalrahman, A. 2010. Green building material requirements and selection (a case study on a HBRC building in Egypt). *ASCE 6th International Engineering and Construction*, (pp. 80–91). Cairo.
Henrikson, R. & Greenberg, D. 2011. Bamboo architecture: in competition and exhibition. Createspace Independent Pub.
Hidalgo Lopez, O. 2003. Bamboo: the gift of the gods. O. Hidalgo-Lopez.
INBAR. 2014. Bamboo: A strategic resource for countries to reduce the effects of climate change. Beijing: *International Network for Bamboo and Rattan.*
INBAR. 2015, June 10. A pathway out of poverty. Retrieved October 11, 2018, from INBAR: https://www.inbar.int/a-pathway-out-of-poverty/.
INBAR. 2017, June 5. Gold is freen: protecting the environemtn. Retrieved November 4, 2018, from INBAR: https://www.inbar.int/gold-is-green-protecting-the-environment-with-bamboo/.
Janssen, J.J. 2000. Technical report no. 20, Designing and building with bamboo. *International Network for Bamboo and Rattan* 2000.
Kahn, L. 1978. Shelter II. Bolinas.
Lala, S.E. 2017. A comparative study on the seismic performance of the different types of bamboo stilt houses of North-East India. J. Environ. Nanotechnol.
Liese, W. 1998. The anatomy of bamboo culms. Brill.
Lobovikov, M.E. 2007. World bamboo resources: a thematic study prepared in the framework of the global forest resources assessment 2005. Food & Agriculture Org.
Lou, Y.E. 2010. Bamboo and Climate Change Mitigation: a comparative analysis of carbon sequestration. *International Network Bamboo and Rattan* (INBAR).
Mahdavi, M.E. 2010. Development of laminated bamboo lumber: review of processing, performance, and economical considerations. *Journal of Materials in Civil Engineering*, 1036–1042.
Mardjono, F. 2002. A bamboo building design decision support tool (Doctoral dissertation). Eindhoven, Netherlands: Fitri Mardjono.
Mohmod, A.L. 1990. Anatomical features and mechanical properties of three Malaysian bamboos. *Journal of Tropical Forest Science*, 227–234.
Olmez, G.E. 2016. The environemntal impacts of iron and steel industry: a life cycle assessment study. *Journal of Cleaner Production*, 195–201.
Rapoport, A. 1969. House form and culture. New Delhi: Prentice-hall of India Private Ltd.
Rodríguez-Camilloni, H. 2009. Rethinking bamboo architecture as a sustainable alternative for developing countries: Juvenal Baracco and Simón Vélez. Proceedings of the Third International Congress on Construction History. Cottbus.
Shim, B. 2010. On site review report: Green School. https://archnet.org/system/publications/contents/8769/original/DTP101268.pdf?1391611188
Sutnaun, S.E. 2011. Macroscopic and microscopic gradient structures of bamboo culms. Walailak *Journal of Science and Technology*, 81–97.
Van Dine, A. 1977. Unconventional Builders. JG Ferguson Publishing Company.
Velez, S.E. 2000. Grow your own house: Simon Velez and bamboo architecture. Vitra Design Museum.
Villegas, M. 2003. New bamboo: architecture and design. Sirsi.
Xiao, Y.E. 2009. Design and construction of modern bamboo bridges. *Journal of Bridge Engineering*, 533–541.
Zoysa, N.E. 1988. Some aspects of bamboo and its utilization in Sri-Lanka. *Bamboos Current Research*, 6–11.
"24H". 2009. Retrieved from https://www.greendotawards.com/submit/upload/2007/large/3-335-11_24H_architecture_Childrens_activity_centre_copy.pdf.

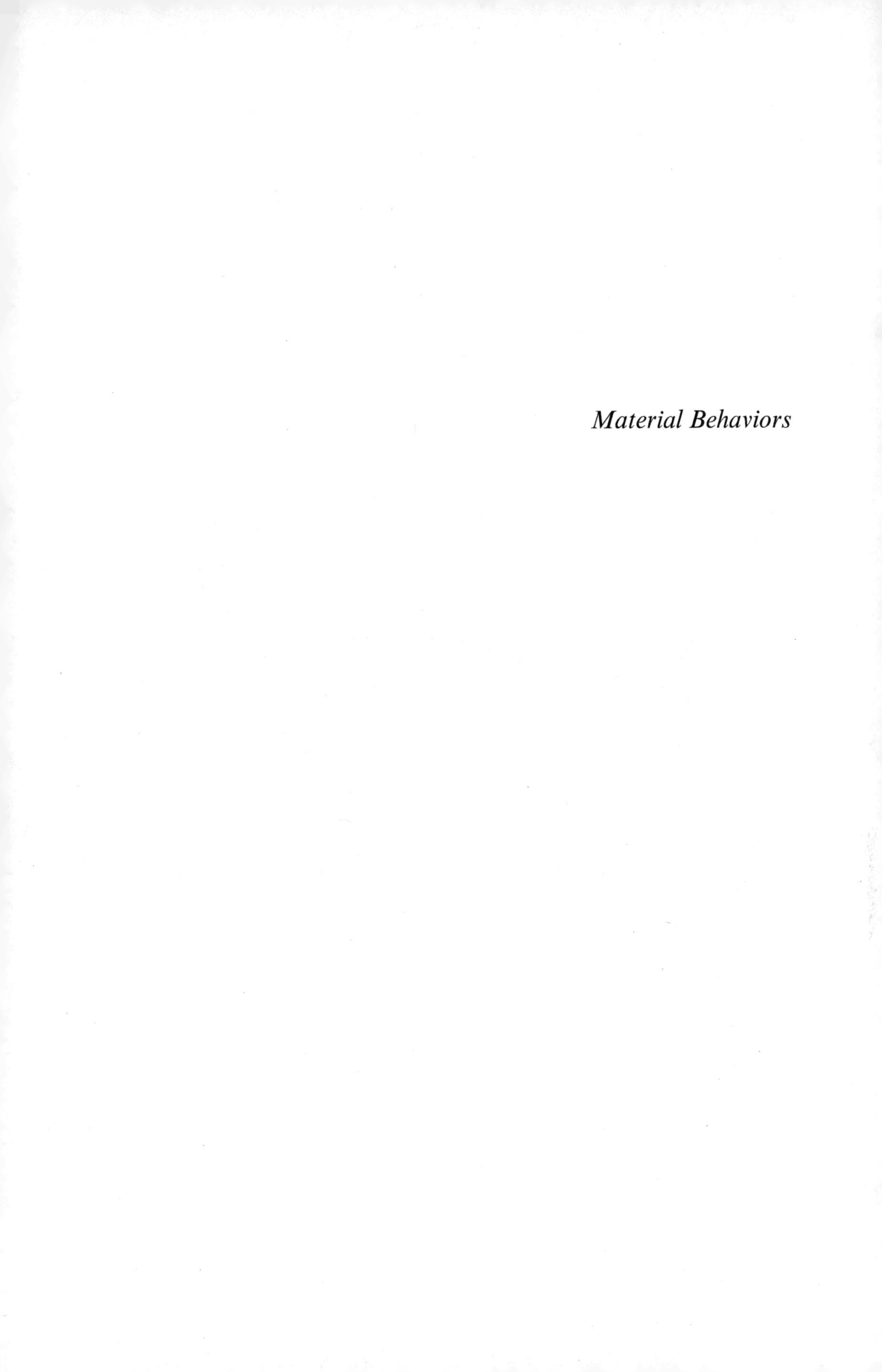

Material Behaviors

Modern Engineered Bamboo Structures – Xiao, Li & Liu (eds)
© 2020 Taylor & Francis Group, London, ISBN 978-1-138-35185-1

Fire-safe bamboo structures – a methodology to facilitate performance-based design

M. Gutierrez, A. Solarte, I. Pope, C. Maluk & J. Hidalgo
The University of Queensland, School of Civil Engineering, St Lucia, Queensland, Australia

J.L. Torero
The University of Maryland, A. James Clark School of Engineering, College Park, Maryland, USA

ABSTRACT: Traditionally, bamboo has only been used for structural purposes in low-rise buildings, and one of the main reasons is that the fire performance of load-bearing bamboo structures is not well understood. Before this material can be used in applications where fire safety considerations are critical, for instance in mid-rise buildings or other complex structures, its fire performance must be characterised comprehensively. This requires an understanding of the fire-induced failure mechanisms and mechanical performance at elevated temperatures of bamboo structural elements, as well as their thermal performance and burning behaviour. The work described herein summarises the knowledge gaps and engineering challenges that must be overcome to facilitate the performance-based design of bamboo structures. A continuing holistic study aimed at addressing these issues is introduced. Key thermal and mechanical properties and burning behaviours of both laminated and round bamboo are summarised.

1 INTRODUCTION

While round bamboo has been widely used in traditional low-rise construction for millennia, engineered bamboo products are novel construction materials that share many of the attractive qualities of engineered timber products, which are becoming increasingly popular for use as a primary structural material in mid- and high-rise buildings. Engineered bamboo structural elements possess a high strength-to-weight ratio and aesthetic value, while reducing the variability and difficulties of connections associated with the natural material (Sharma et al. 2014). Furthermore, bamboo has significant potential for sequestering carbon, and may often be more readily available than timber in regions with large populations and developing economies, including Latin America, Africa, and East Asia (Yiping et al. 2010). As a combustible material, the use of bamboo in construction has thus far been constrained to secondary applications or low-rise buildings, due to a lack of research and understanding of its fire performance.

To ensure life safety in building fires, tenable conditions must be maintained along egress routes until all occupants have reached a safe space. Additionally, the structure must be resilient to the effects of the fire throughout egress and firefighter intervention, and as appropriate to prevent damage to neighbouring property. For tall buildings with very long egress times, this second requirement effectively means that the structure must survive the whole duration of the fire until burnout and through the cooling period. However, occupants of low-rise buildings may typically exit the building before the fire has grown sufficiently to begin compromising the structure, and firefighters can often intervene externally when necessary.

For a lignocellulosic material such as bamboo, the question of fire safety is therefore dependent on a series of interrelated factors, whose relative significance is influenced by the context and scale at which the material is used in a building (Reszka & Torero 2016). The chemical and thermal properties of the material – as well as its use and configuration in the building – govern the parameters of flammability, flame spread, and heat release rate, which determine its contribution to the growth of the fire. These affect the development of untenable conditions within the building, and therefore the available time for occupants to egress safely, but potentially also the thermal loading on the structure. Thermal properties and burning behaviour, such as charring, also dictate the heat-transfer and development of temperature profiles through structural elements. This feeds back into the dynamics of the fire, but also the performance of the structure, since the mechanical properties of bamboo have been shown to be greatly affected by elevated temperature (Gutierrez Gonzalez et al. 2018, Xu et al. 2017, Mena et al. 2012). Therefore, in order to ensure the fire-safe design of bamboo structures it is essential to analyse all of these factors holistically, and with due consideration of the context of the building and its occupants. This document will outline the research methodology required to address most of these factors, and to provide tools for engineers and practitioners to understand how to design fire-safe bamboo structures.

2 MATERIAL PROPERTIES AND FLAMMABILITY

The characterisation of bamboo's performance under fire conditions requires fundamental knowledge of the thermal degradation and flammability properties of its components, because these will determine the onset of hazard. With this understanding, a methodology could be established to determine how the material affects the development of the fire and the safety of occupants within a building. The comprehension of bamboo's physical, thermal, and flammability behaviour properties is of key importance, and serves as a starting point for analysis.

2.1 *Thermal characterisation*

Woody bamboos are mainly composed of polymers that undergo different transformations when exposed to heat. The anatomical structure of bamboo can be described as a two-phase composite of vascular bundles and parenchyma tissue (Amada et al. 1997). The vascular bundles are mainly made of fibre bundles and vessels, and the parenchyma tissue is a porous media in which the fibre bundles are embedded (Liese 1998). The bamboo fibres are made of approximately 73.8% cellulose, 10.1% lignin and 12.5% hemicellulose (Wang et al. 2009). The parenchyma cells have a relatively similar content of cellulose (68.8%), but it is much less lignified than the fibre bundles and it is characterised by having thinner cell walls, large cell cavities, and larger and more pits compared to bamboo fibre bundles (Wang et al. 2015). When subjected to heating, these polymers can experience a chemical decomposition, also known as pyrolysis. During pyrolysis the bamboo will yield products resulting in a mixture of volatile gases, called pyrolysates, and in reaction with oxygen can sustain flaming combustion (Drysdale 2011). Therefore, it can be said that the onset of hazard for flame spread and fire growth is when the material reaches a pyrolysis temperature that will enable ignition.

Thermogravimetric analysis of bamboo conducted in oxygen shows three distinct phases. For temperatures between ambient and 150°C, the material suffers some mass loss due to the evaporation of water within the cellular walls and cavities (free and hygroscopic water). At temperatures above 150°C and below 400°C, the material experiences the highest loss of mass. Around 165°C, the derivative of mass loss from thermogravimetric data (DTG) shows that the mass loss rate of bamboo starts to increase significantly. This can be identified as the point of the initiation of pyrolysis until the hemicellulose and cellulose reach the peak of volatilisation at 287°C and 333°C respectively, where the maximum thermal degradation is observed (Chen et al. 2015, Jiang et al. 2012, Wang et al. 2008). The DTG starts to decrease after this, and at 400°C it can be observed that bamboo has already lost about 70% of its mass. At temperatures above 400°C, the material continues losing mass at a lower rate thanks to the oxidation of the char, until only the inorganic material remains, and the TGA shows approximately 20-25% of residual mass at 900°C (Mi et al. 2016, Jiang et al. 2012).

Another key parameter to understand the fire performance of bamboo structures is to characterise the heat transfer mechanisms of this material. To do this, the thermal physical properties –

density (ρ), thermal conductivity (k) and specific heat capacity (C_p) – and thermodynamic properties of bamboo – thermal inertia ($k\rho C_p$) and thermal diffusivity (α) – need to be identified.

Compared to engineered timber products such as Glulam, engineered bamboo products have a higher density (Sharma et al. 2015), and a higher thermal conductivity (Solarte et al. 2018). The thermal conductivity is a property that drives a material's ability to conduct heat, and it determines the rate at which energy is transported by the diffusion process. A material with a high thermal conductivity and density will result in a material with high thermal inertia. The thermal inertia is the product of the thermal conductivity, the density and the specific heat capacity ($k\rho C_p$), and is the resistance of a material to change its surface temperature. In other words, it measures how quickly the surface temperature of a material increases when exposed to heat. Materials with high values of thermal inertia require higher energy or heat flux to increase their surface temperature, so in comparison with timber, laminated bamboo products will require more energy to ignite. Thermal diffusivity (α) is another important property to identify, and it is calculated from the ratio between the thermal conductivity and the product of density and specific heat. This measures the ability of a material to conduct thermal energy relative to its ability to store it. The larger the thermal diffusivity, the faster the temperature will increase at a certain depth in a material (Drysdale 2011). In comparison with engineered timber products, the heat wave may penetrate in a faster manner through bamboo structures. Hence elevated temperatures inside a structural element could reach farther than in timber.

The characterisation of the thermal degradation of bamboo structures is important for design, because even if at this stage the material has not ignited, it does not necessarily mean that the building is fire safe. Even when there is no flaming combustion, pyrolysis can produce toxic gases that can affect people before evacuation. Before the onset of ignition, the structure may also lose integrity due to phase changes (e.g. glass transition) enabled by the transfer of heat through the cross-section.

2.2 *Flammability*

Once the thermal properties have been characterised, and pyrolysis temperatures have been identified, the next step to design fire safe bamboo buildings is to identify the conditions in which ignition takes place. The quantification of flammability in materials is fundamental because it determines the general fire behaviour, size and fire spread, which all control the development of untenable conditions within a building, i.e. conditions that are hazardous to the health and safety of occupants.

To determine the ignition conditions of bamboo, two main steps determine the characterisation of flammability for this methodology. First, obtaining parameters such as the critical heat flux for piloted ignition (q_{cr}"), and time and temperature for ignition (t_{ig}, T_{ig}) as stated by Torero (Torero 2013). The critical heat flux for ignition is the highest heat flux at which ignition does not take place, i.e. the material will ignite at any higher heat flux than this. This is obtained experimentally by testing samples and obtaining the time it takes to ignite at various heat fluxes. (Torero 2013). Secondly, analysing the burning process of the material by determining the critical mass loss rate at the initiation of sustained flaming combustion (Solarte et al. 2018, Rasbash et al. 1986).

Another flammability aspect to consider for design is how the flame will spread and contribute to fire growth inside a compartment or façade made of bamboo. This is a mechanism in which a flame moves forward in the vicinity of a pyrolysing region of the surface fuel (Fernadez-Pello et al. 1983). There are two major phenomena that need to be characterised in the understanding of flame spread: (1) opposed flow flame spread, and (2) concurrent flow flame spread (Fernadez-Pello et al. 1983). Opposed flow flame spread addresses the scenario in which the flame front is moving forward against the flow of air, only preheating the unburned fuel immediately ahead of the flame front. Upward vertical flame spread is an example of concurrent flow flaming, and in this case the flow is pushing in the same direction as the flame, effectively "driving" it forward and enabling faster propagation by preheating the material further ahead of the flame.

In the event of a fire, a safe evacuation needs to be guaranteed. This means that every person should have enough time to egress before untenable conditions are reached. Untenable conditions may be caused by reduced visibility and elevated toxicity due to smoke, or high temperatures and heat fluxes within the building, all of which are intrinsically linked to the growth and spread of the fire. Once the flame spread parameters of the material are established, the rate at which the fire will grow and contribute to the production of heat and smoke may be estimated. By understanding this behaviour, it is possible to design a bamboo building in which the development of untenable conditions will not occur until after the safe evacuation of the building.

3 THERMAL BEHAVIOUR

The burning behaviour of bamboo, like timber, is characterised by a series of chemical and physical changes occurring over a wide range of temperatures. As detailed in Section 2.1, these changes are related to the different components of the material, which are predominantly lignin, cellulose, hemicellulose, and water. As the temperature increases, the first effect is the evaporation and migration of moisture, which occurs rapidly around the boiling temperature of 100°C. This is effectively an endothermic process, due to the extra energy required to evaporate water, but the migrating moisture may also transfer heat through the material. Above 100°C, all of the free water has been driven out, and the remaining components begin to chemically decompose through the process of pyrolysis (Bartlett et al. 2018b). This peaks in a temperature range of around 200-400°C (Chen et al. 2015, Jiang et al. 2012), as the cellulose, hemicellulose and lignin volatilise and are converted into char, which is primarily composed of carbon. As temperatures rise further, exothermic oxidation reactions take place within the char, providing an additional heat source and resulting in gradual regression of the char layer (Bartlett et al. 2018b).

In a structural member under transient heating, these processes occur over different temperature zones that move through the section as the heat wave travels from the exposed surface. As a simplification, these regions are often delineated by isotherms that correspond to the critical temperatures for each process. Typically, the region below 100°C is assumed to possess the thermal properties – density, conductivity and heat capacity – of the virgin material. Between this 'normal zone' and the 300°C isotherm is the 'dry zone', from which all of the moisture has been removed and the thermal properties are altered accordingly. Above 300°C, the material is commonly assumed to have the thermal properties of char, which typically has much lower density and conductivity, so acts as an insulating barrier protecting the material beneath. However, charring is associated with shrinkage and thermal stresses that induce cracking of the charred material, locally compromising this insulation (Reszka & Torero 2016). It should be noted that these critical temperatures are approximations, and in reality the various processes will occur over temperature ranges that are dependent upon the heating rate and the exact chemistry of the material. Therefore, the boundaries of these temperature zones should be experimentally validated before they can be used for design purposes. Similarly, the density, conductivity and heat capacity are key determinants of heat transfer through solids, so it is essential to characterise these values over the different regions in order to predict how temperature profiles will develop in-depth during a fire.

3.1 *Implications for structural fire performance models*

For round bamboo culms, with wall thicknesses of typically only 10-20 mm, structural members exposed to fire conditions are likely to reach critical failure temperatures rapidly, particularly since the fibres that contribute most to loadbearing capacity are concentrated near the surface. In this scenario, coupled thermal and structural models are likely to be redundant, since the exposed structural elements might conservatively be expected to fail shortly after the fire has developed. However, where round bamboo elements are protected by other insulating components – as may be the case in stud wall systems such as 'bahareque' structures – the heating rate will be lower, and transient thermal models may provide a useful input for predictions of structural degradation. This would also require modelling heat flow through the insulating elements, and analysis of the whole system.

For 'engineered' bamboo products, such as laminated bamboo or bamboo scrimber, the cross-sections of structural elements can be made much thicker, allowing for an insulating char layer to form while some structural capacity is maintained by the inner material. In this case, the development of elevated temperatures ahead of the charring front may become significant in determining the residual structural capacity. Another characteristic of engineered bamboo products is the role of adhesives in binding together the bamboo components into a composite material. Whether in the form of discrete adhesive layers between bamboo strips, or an amorphous resin matrix binding bamboo fibre bundles, as is the case for products like bamboo scrimber, the performance of the glue is critical to the overall behaviour of the structure. Many common structural adhesives begin to degrade at temperatures as low as 100-300°C (Clauß et al. 2011), potentially resulting in delamination of the outer lamellae or a premature loss of strength under fire conditions. Delamination of the outer bamboo layers can totally compromise the insulating effect of the char layer, allowing additional heat to reach the virgin material. A necessary condition for a bamboo structure to survive a fully developed fire is that it will eventually stop burning of its own accord, once the other combustible contents have been consumed by the fire. This process, called 'self-extinction', is a result of the increasing char layer limiting the amount of energy that can pyrolyse the unburnt material, until it is below the minimum required to sustain burning. Progressive delamination can critically undermine the self-extinction process, so it is vital to be able to predict if or when each glue layer might reach its failure temperature in order to assess this potential.

A standard approach for timber structures in fire is to assume a nominal charring rate of around 0.6-0.7 mm/min to approximate the movement of the charring front, ahead of which is an assumed 'heated depth' of 7 mm (CEN 2004). This depth is assumed to account for the reduced strength of the material heated above ambient but below the charring temperature, and the material in this region is assumed to have no structural capacity. By subtracting the assumed zero-strength char layer and heated depth, the residual capacity of the structural member is calculated for a particular fire exposure time. While it may seem reasonable to apply this approach to engineered bamboo structures, due to the chemical similarities of timber and bamboo, it has been found that this method cannot predict structural performance at certain fire exposures (Bartlett et al. 2018a). This has been attributed to a much larger heated depth observed in laminated bamboo fire experiments, which is likely due to the fact that bamboo has substantially different thermal properties than the timbers for which this method has been developed. In particular, the thermal conductivity of laminated bamboo has been found to be approximately double that of timber (Solarte et al. 2018, Bartlett et al. 2018a), which results in faster in-depth heating.

Therefore, the performance-based design of bamboo structures requires a specific understanding of how the materials behave under fire conditions, and experimentally validated thermal models to predict the in-depth temperature rise under a range of heating conditions. These predictions can then inform tailored structural models that account for the unique material behaviour at elevated temperatures.

4 STRUCTURAL FIRE PERFORMANCE

4.1 *Thermal degradation of bamboo*

Normally, the rise of temperatures can lead to several chemical and physical transformations in a material, and in most of those cases, these changes immediately compromise the strength and load-bearing capacity. As it was already mentioned, bamboo compounds experience thermal degradation at different temperatures affecting their strength, and stiffness. Before the mass loss due to evaporation of water, several authors have reported that lignin suffers a significant loss of stiffness at around 60-70°C, based on previous studies of the dynamic mechanical analysis (DMA) of bamboo (Reszka and Torero 2016, Liu et al.

2012, Ramage et al. 2017). It is well known that one of the methods used to bend bamboo poles utilises localised high-temperature to soften the bamboo cross-section, allowing to bend the pole into different shapes through the use of jigs (Hidalgo López 2003). Additionally, it has been reported that when thermal modification is applied to engineered bamboo products, a softening of the bamboo cross-section is achieved. When bamboo is treated with temperatures of around 150°C, and then is pressed in the radial direction, it is possible to increase its density due to a reduction of voids and a redistribution of fibres. This will also increase the mechanical properties up to 30% (Archila Santos et al. 2014, Yang et al. 2016), however, if the temperature applied is above 150°C, cellulose and hemicellulose will decompose, and the modulus of rupture and the modulus of elasticity will drop compared to the values at ambient temperature (Zhang et al. 2013, Trujillo and López 2016).

A comprehensive thermo-chemical characterisation should be conducted to understand the degradation of the different compounds of bamboo, since the fibre bundles and the parenchyma tissue have different effects on the load-bearing capacity of bamboo. Whereas the fibres are mainly responsible for the tensile strength, the parenchyma plays a fundamental role in supporting the fibre bundles under compression, besides giving the strength of bamboo under shear stress parallel to the fibres (Rizal et al. 2018).

TGA results of the bamboo fibres and matrix can explain the thermal degradation and the temperature range at which pyrolysis occurs, identifying mass loss and other phenomena affecting bamboo strength, like the progression of the char front and the reduction of the cross-section. DMA analysis of the bamboo fibres and matrix will help to explain the degradation in the stiffness at different temperatures, and how the material may undergo processes that, even without producing mass loss or pyrolysis reactions, are able to modify the mechanical behaviour, resulting in significant strength losses under different load scenarios. Results from this characterisation will help to understand the reduction of the mechanical properties of the whole bamboo cross section.

4.2 *Mechanical properties of bamboo at elevated temperature*

Several authors have suggested methods to measure the reduction of the mechanical properties of structural materials at elevated temperatures, however Pettersson (1986) suggested that the mechanical behaviour at elevated temperatures can be assessed through either steady state or transient-state tests. Under steady state tests, the cross-section of the load-bearing element is at a constant temperature during testing, and therefore, the test is not dependent on time once the steady state has been reached (Pettersson 1986). At this temperature, tests to measure stress-strain relationships can be conducted under load-controlled or strain-controlled rate, as well as other tests at constant stress or constant strain where creep or relaxation can be obtained (Pettersson 1986, Purkiss and Li 2014, Buchanan and Abu 2017).

As with other lignocellulosic materials, bamboo suffers a significant reduction of the mechanical properties at elevated temperatures. Under compression, bamboo's strength is provided by the composite action of fibre and parenchyma (Sharma et al. 2015). The parenchyma tissue gives the fibres lateral support, avoiding the buckling failure of individual fibres. On the other hand, fibres are the primary source of tensile strength in bamboo (Gutierrez Gonzalez et al. 2012). Results from Xu et al. (2017) showed that bamboo scrimber suffers a significant reduction in the mechanical properties at temperatures below 200°C. Xu reported that bamboo scrimber samples under compression showed a strength reduction of 55% at 100°C, whereas the tensile strength reduction at 100°C is around 38% (Xu et al. 2017). Mena et al. (2012) showed that round and laminated bamboo suffer a reduction in bending strength of approximately 8% and 18% respectively at 100°C (Mena et al. 2012).

Similar to timber, bamboo scrimber experiences a lower reduction of the modulus of elasticity (MOE) at elevated temperatures compared to the reduction in the compressive and tensile strength. At 100°C, the MOE of bamboo scrimber measured in compression presented a reduction of 20%, whereas in tension the reduction is about 12%.

Despite the indications of these studies regarding the reduction in the mechanical properties of bamboo, there have as yet been no studies reporting the reduction in the compressive, tensile, and shear strength of round and laminated bamboo at elevated temperatures. Thanks to the use of novel set-ups, the reduction of the mechanical properties of round and laminated bamboo is being investigated at The University of Queensland using an environmental chamber, and a heating blanket set-up. The studies will be conducted with samples prepared following the recommendations of ISO 22157 for round bamboo (ISO 2004a, ISO 2004b), and EN 408 for laminated bamboo (European Committee for Standardization 2012), refer to Figure 1. However, some parameters had to be changed due to the conditions for testing at elevated temperatures, or due to the unique conditions of bamboo, which are not addressed in the timber standard. Results from these tests will provide the required information to create constitutive models of bamboo at elevated temperature, and will enable the prediction of the mechanical behaviour of load-bearing elements under transient heating.

4.3 *Load-bearing capacity of bamboo elements under fire*

The understanding of the reduction in the mechanical properties of bamboo at elevated temperature will enable the prediction of the mechanical response of load-bearing elements under different load conditions. For instance, when heat is applied on one side of the element, the combustion of bamboo can lead to charring of the exposed surface. The reduction in the mechanical properties indicates that above a certain temperature, the material cannot withstand load anymore, and it will lose its mechanical resistance (Buchanan and Abu 2017, Buchanan et al. 2014). For temperatures above that value, it can be assumed that the material strength and stiffness is zero, and therefore further analysis should be conducted to understand the correlation between mechanical properties and the temperature gradient over the cross-sectional area, especially where temperatures are between ambient and the temperature at which the strength is assumed to be zero. This analysis can lead to different mechanical behaviours, such as the loss of some parts of the cross-section in certain areas exposed to heat, or to a shift in the neutral axis of the cross-section due to the variation of mechanical properties along the depth.

The capacity to predict the temperature profiles/gradients and the constitutive models of bamboo at elevated temperature is a fundamental input to predict the load-bearing response,

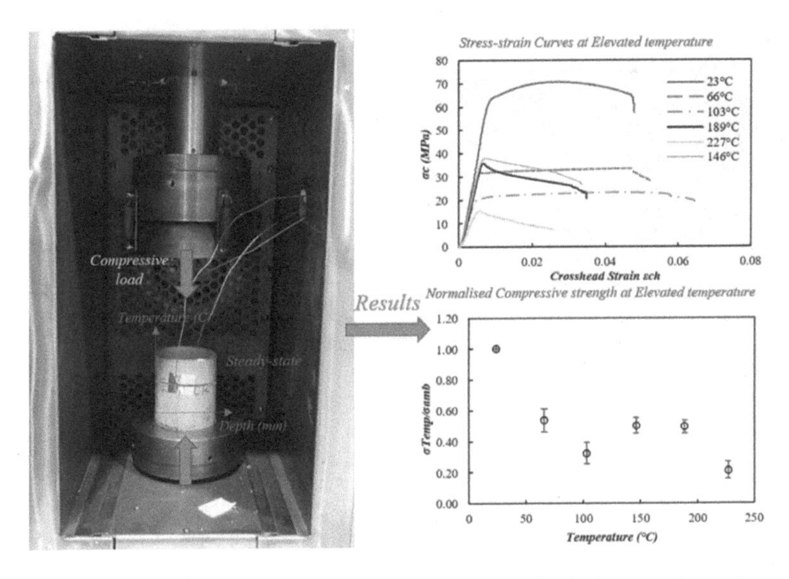

Figure 1. Experimental set-up to measure the reduction in the mechanical properties at elevated temperatures at steady state.

as is explained in Figure 2. The analysis of deformations, strains, and stress-states during a fire will allow predicting failure loads, displacements and times to failure of single elements, or even structural systems. These design outputs are required for compatibility with other outputs from the fire safety strategy developed to guarantee the safety of occupants, particularly when structural integrity must be provided for the safe evacuation of the building, or for the safety of the fire brigades controlling the fire in a bamboo structure (Östman et al. 2017).

Bearing this in mind, laminated bamboo has shown the capacity to withstand loads under fire for longer periods than round bamboo. Due to the geometry of round bamboo, which normally possesses a small wall thickness, combined with its already reported thermal properties, round bamboo load-bearing elements lose cross-section rapidly, reaching failure loads in short time-periods (Kaminski et al. 2016). Tests on round bamboo have shown that thermal gradients are hard to achieve, therefore it is not practical to measure the reduction in the mechanical properties along the cross-section (Gutierrez Gonzalez et al. 2018). The progression of the char front is observed, and the load-bearing capacity of round bamboo elements can be lost rapidly during a typical compartment fire. Understanding the burning behaviour and the structural performance of round bamboo structures during a fire will define the limits of applicability of this material in the urban environment. However, due to the geometry of laminated bamboo, this material is able to create char layers that may provide enough insulation to the inner parts of the cross-section, having residual cross-sections with intact mechanical properties that could help to withstand design loads through the duration of a fire. After understanding the behaviour of laminated bamboo under a fire, this material could have the potential to be used in applications where structural fire safety considerations must be addressed, as has been already seen in high-rise timber buildings made of engineered wood products. Mechanical tests during fire should be conducted to guarantee that loads, deformations, and time to failure could be accurately predicted when designing load-bearing laminated bamboo elements. This will be the basis for understanding the behaviour of bamboo structural systems under fire.

Thermal gradients within the cross-section are also key to predicting other phenomena, like delamination. The reduction in the mechanical properties of the glue at elevated temperatures, as well as the shear stress in the element at the glue line location, will determine whether or not delamination can occur at any point. Delamination is not only important to maintain a certain level of strength in the load-bearing element, but it is also an important parameter to guarantee the conditions of self-extinguishment.

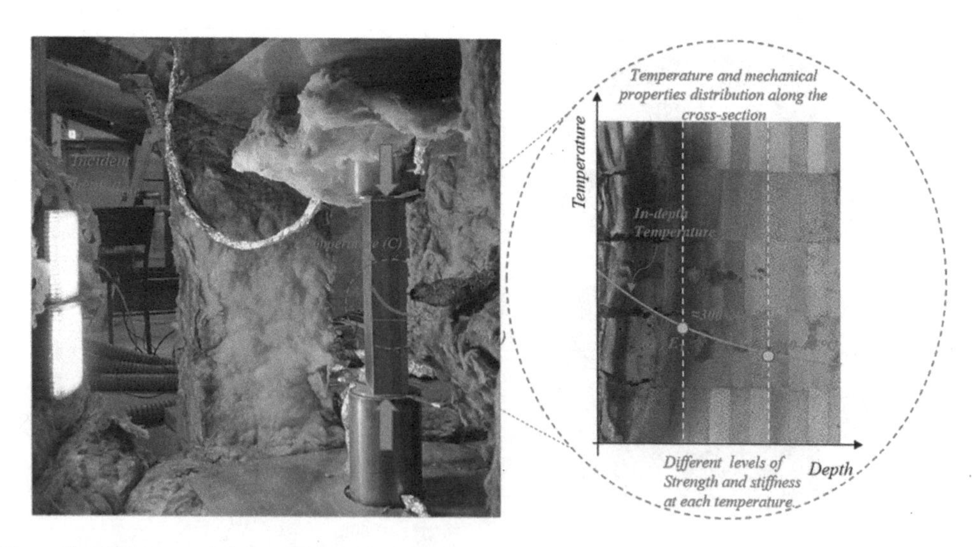

Figure 2. Diagram of the correlation between the thermal behaviour and the mechanical properties of bamboo at elevated temperatures.

4.4 *Implications of the thermo-mechanical degradation of bamboo in the built environment*

The reduction of the mechanical properties at different temperatures could be included in building codes that require the prediction of the load-bearing capacity of bamboo under fire. Current bamboo codes do not allow building with exposed bamboo elements due to lack of understanding of how load-bearing capacity of bamboo elements can be predicted under exposure to different fire scenarios (ISO 2004c). If designers had the information required to calculate the temperature profile and the reduction in mechanical properties within the bamboo cross-section under any fire scenario, fire and structural engineers would be able to predict the failure of load-bearing elements during a fire. Current prescriptive design frameworks have established different fire resistance levels to prescribe a minimum structural performance within the scope of the whole fire safety strategy. However, these frameworks were created for other conventional materials like concrete and steel. Since these are non-combustible in nature, the current design guidelines will not describe the structural performance of bamboo in an appropriate manner. New design guidelines should be proposed to analyse the structural performance of combustible materials like bamboo, where real fires should be used as the imposed thermal boundary conditions, and where the analysis of their constitutive models at elevated temperatures will be the basis to predict their mechanical response under a fire.

5 CONCLUSIONS

To achieve fire safe bamboo structures, it is important to develop new tools that enable a performance-based methodology and take into consideration the complete characterisation of the material, such as the thermal degradation, flammability conditions, heat transfer, and understanding how all of this can affect the structural behaviour of the material and the development of the fire.

The understanding of the intrinsic characteristics of bamboo will determine the different failure modes associated with the structural failure for each specific scenario. This full characterisation will also enable professionals to design with the knowledge required to define realistic fire scenarios, which is fundamental to guarantee that tenable conditions are maintained in the building until occupant egress is complete.

While it is tempting to assume that standard design methodologies used for timber can be applied equally to bamboo products, recent studies have shown that this may not be the case, and that new approaches are required, tailored to the specific materials. This will involve a detailed characterisation of the thermal response of bamboo structural elements under realistic fire conditions so that temperature profiles through the depth of the cross-section can be predicted and used as an input in structural models.

Correct and appropriate testing methodologies and very well-defined thermal boundary conditions are key aspects to assess the reduction in the mechanical properties at elevated temperatures.

The understanding of the reduction of the mechanical properties at elevated temperatures will enable the prediction of the cross-section reduction due to the fire, and it will allow predicting the loss of load-bearing capacity, as well as different failure mechanisms caused by thermal loading.

The understanding of the structural performance of bamboo structures during a fire will allow the bamboo industry to build structures where the fire safety strategy requires guaranteeing structural integrity before and after fire service intervention. A good understanding of the behaviour of bamboo at elevated temperatures, will unlock the capacity of bamboo to be used as a mainstream construction material in urban environments.

REFERENCES

Amada, S., Ichikawa, Y., Munekata, T., Nagase, Y. & Shimizu, H. 1997. Fiber texture and mechanical graded structure of bamboo. *Composites Part B: Engineering*, 28, 13–20.

Archila Santos, H.F., Ansell, M.P. & Walker, P. 2014. Elastic properties of thermo-hydro-mechanically modified bamboo (Guadua angustifolia Kunth) measured in tension, *Key Engineering Materials*, 600: 111-120.
Bartlett, A.I., Chapman, A., Roberts, A., Wiesner, F., Hadden, R.M. & Bisby, L.A. 2018a. Thermal and flexural behaviour of laminated bamboo exposed to severe radiant heating. *2018 World Conference on Timber Engineering (WCTE 2018), 2018: Seoul, Korea.*
Bartlett, A.I., Hadden, R.M. & Bisby, L.A. 2018b. A Review of Factors Affecting the Burning Behaviour of Wood for Application to Tall Timber Construction. *Fire Technology*, 1–49.
Buchanan, A., Ostman, B. & Frangi, A. 2014. *Fire resistance of timber structures*, National Institute of Standards and Technology Gaithersburg.
Buchanan, A.H. & Abu, A. 2017. *Structural design for fire safety/Andrew H. Buchanan, Anthony K. Abu*, Chichester: John Wiley & Sons Inc.
CEN 2004. Eurocode 5: Design of timber structures - Part 1-2: General - Structural fire design (EN 1995-1-2). Brussels: European Committee for Standardization.
CEN 2012. Timber structures - Structural timber and glued laminated timber - Determination of some physical and mechanical properties (EN 408:2010+A1). Brussels: European Committee for Standardization.
Chen, D., Liu, D., Zhang, H., Chen, Y. & Li, Q. 2015. Bamboo pyrolysis using TG–FTIR and a lab-scale reactor: Analysis of pyrolysis behavior, product properties, and carbon and energy yields. *Fuel*, 148: 79–86.
Clauß, S., Joscak, M. & Niemz, P. 2011. Thermal stability of glued qood joints measured by shear tests. *European Journal of Wood and Wood Products*, 69(1): 101–111.
Drysdale, D. 2011. *An introduction to fire dynamics*. Chichester: John Wiley & Sons Inc.
Fernandez-Pello, A. & Hirano, T. 1983. Controlling mechanisms of flame spread. *Combustion Science and Technology*, 32(1-4): 1–31.
Gutierrez Gonzalez, M., Madden, J. & Maluk, C. 2018. Experimental Study on Compressive and Tensile Strength of Bamboo at Elevated Temperatures. *2018 World Conference on Timber Engineering (WCTE 2018), 2018: Seoul, Korea.*
Gutierrez Gonzalez, M., Takeuchi, C. & Perozo, M.C. 2012. Variation of tensile strength parallel to the fiber of Bamboo Guadua angustifolia kunth in function of moisture content. *Key Engineering Materials*, 517: 71–75.
Hidalgo López, O. 2003. *Bamboo: The Gift of the Gods.*
ISO, I.O.F.S. 2004a. Bamboo – Determination of physical and mechanical properties – Part 1: Requirements (ISO 22157-1:2004). Geneva: ISO.
ISO, I.O.F.S. 2004b. Bamboo – Determination of physical and mechanical properties – Part 2: Laboratory manual (ISO/TR 22157-2:2004). Geneva: ISO.
ISO, I.O.F.S. 2004c. Bamboo – Structural design (ISO 22156:2004). Geneva: ISO.
Jiang, Z., Liu, Z., Fei, B., Cai, Z., Yu, Y. & Liu, X.E. 2012. The pyrolysis characteristics of moso bamboo. *Journal of Analytical and Applied Pyrolysis*, 94: 48–52.
Kaminski, S., Lawrence, A. & Trujillo, D. 2016. Design Guide for Engineered Bahareque Housing. Beijing: International Network for Bamboo and Rattan (INBAR).
Liese, W. 1998. The Anatomy of Bamboo Culms. INBAR Technical Report N° 18. In: International Network for Bamboo And Rattan, I. (ed.). Beijing: International Network for Bamboo and Rattan (INBAR).
Liu, Z., Jiang, Z., Cai, Z., Fei, B., Yu, Y. & Liu, X. 2012. Dynamic Mechanical Thermal Analysis of Moso Bamboo (Phyllostachys heterocycla) at Different Moisture Content. *Bioresources*, 7(2): 1548–1557.
Mena, J., Vera, S., Correal, J.F. & Lopez, M. 2012. Assessment of fire reaction and fire resistance of Guadua angustifolia kunth bamboo. *Construction and Building Materials*, 27: 60–65.
Mi, B., Liu, Z., Hu, W., Wei, P., Jiang, Z. & Fei, B. 2016. Investigating pyrolysis and combustion characteristics of torrefied bamboo, torrefied wood and their blends. *Bioresource Technology*, 209: 50–55.
Östman, B., Brandon, D. & Frantzich, H. 2017. Fire safety engineering in timber buildings. *Fire Safety Journal*, 91: 11–20.
Pettersson, O. 1986. Structural fire behaviour - development trends. *LUTVDG/TVBB–3031–SE*. Lund, Sweden: Department of Fire Safety Engineering and Systems Safety, Lund University.
Purkiss, J.A. & Li, L.-Y. 2014. *Fire safety engineering design of structures John A. Purkiss and Long-Yuan Li*, Boca Raton, Fla.: CRC Press.
Ramage, M.H., Sharma, B., Shah, D.U. & Reynolds, T P.S. 2017. Thermal relaxation of laminated bamboo for folded shells. *Materials & Design*, 132: 582–589.
Rasbash, D., Drysdale, D. & Deepak, D., Critical heat and mass transfer at pilot ignition and extinction of a material. *Fire Safety Journal*, 1986. 10(1): 1–10.

Reszka, P. & Torero, J.L. 2016. Fire Behaviour of Timber and Lignocellulose. In Belgacem, N. & Pizzi, A. (eds), *Lignocellulosic Fibers and Wood Handbook*: 555–581. Scrivener Publishing LLC.
Rizal, S., Shawkataly, A.K.H.P., Ikramullah, Bhat, I.U.H., Huzni, S., Thalib, S., Mustapha, A. & Saurabh, C.K. 2018. Recent Advancement in Physico-Mechanical and Thermal Studies of Bamboo and its Fibers. In: Khalil, H.P.S.A. (ed.), *Bamboo: Current and Future Prospects*: 145–164. IntechOpen.
Sharma, B., Gatoo, A., Bock, M. & Ramage, M. 2015. Engineered bamboo for structural applications. *Construction and Building Materials* 81: 66–73.
Solarte, A., Hidalgo, J.P. & Torero, J.L. 2018. Flammability Studies for the Design of Fire-Safe Bamboo Structures. *2018 World Conference on Timber Engineering (WCTE 2018), 2018: Seoul, Korea.*
Torero, J. 2016. Flaming ignition of solid fuels. In Hurley, M.J. et al. (eds), *SFPE Handbook of Fire Protection Engineering*: 633–661. New York: Springer.
Trujillo, D. & López, L.F. 2016. 13 - Bamboo material characterisation. In Harries, K.A. & Sharma, B. (eds), *Nonconventional and Vernacular Construction Materials*: 365–392. Elsevier Ltd.
Wang, G., Li, W., Li, B. & Chen, H. 2008. TG study on pyrolysis of biomass and its three components under syngas. *Fuel*, 87: 552–558.
Wang, H., Zhang, X., Jiang, Z., Li, W. & Yu, Y. 2015. A comparison study on the preparation of nanocellulose fibrils from fibers and parenchymal cells in bamboo (Phyllostachys pubescens). *Industrial Crops and Products*, 71: 80–88.
Wang, Y., Wang, G., Cheng, H., Tian, G., Liu, Z., Xiao Qun, F., Zhou, X., Han, X. & Gao, X. 2009. Structures of Bamboo Fiber for Textiles. *Textile Research Journal*, 80: 334–343.
Xu, M., Cui, Z., Chen, Z. & Xiang, J. 2017. Experimental study on compressive and tensile properties of a bamboo scrimber at elevated temperatures. *Construction and Building Materials*, 151: 732–741.
Yang, T.-H., Lee, C.-H., Lee, C.-J. & Cheng, Y.-W. 2016. Effects of different thermal modification media on physical and mechanical properties of moso bamboo. *Construction and Building Materials*, 119: 251–259.
Yiping, L., Yanxia, L., Buckingham, K., Henley G. & Guomo Z. 2010. *Technical Report 32: Bamboo and Climate Change Mitigation.* Beijing: International Network for Bamboo and Rattan (INBAR).
Zhang, Y.M., Yu, Y.L. & Yu, W.J. 2013. Effect of thermal treatment on the physical and mechanical properties of phyllostachys pubescen bamboo. *European Journal of Wood and Wood Products*, 71: 61–67.

Modern Engineered Bamboo Structures – Xiao, Li & Liu (eds)
© 2020 Taylor & Francis Group, London, ISBN 978-1-138-35185-1

Experimental and numerical assessment of the thermal insulation performances of bamboo shear walls

C. Demartino
Nanjing Tech University, Nanjing, Jiangsu, China

Y. Xiao
Zhejiang University-University of Illinois at Urbana Champaign Institute, Jiaxing, Zhejiang, China
Nanjing Tech University, Nanjing, Jiangsu, China
Department of Civil Engineering, University of Southern California, Los Angeles, CA, USA

J.S. Wang
Guangdong Provincial Academy of Building Research Group Co., Ltd., Guangzhou, Guangdong, China

ABSTRACT: Four light frame shear walls representing one wood-based wall, one hybrid bamboo/wood-based wall and two full bamboo-based walls were studied by using a guard hot box testing apparatus to obtain the thermal resistance and thermal transmittance. The method proposed by ISO 6946 and the finite element method were used to analyze the thermal insulation performance of the light frame walls. The reliability of these two simulation methods was verified by comparison between them and with the experimental results. The results show that the thermal insulation performance of light frame bamboo and wood walls are similar allowing for the feasible substitution of wood by bamboo in real light frame constructions.

1 INTRODUCTION

The reduction of energy losses in buildings is of utmost importance in order to ensure the sustainability and the reduction of heating costs. The energy-loss of walls is around 35%, the door is 15%, the window is 10%, the floor is 15%, and the roof is 25% in the light frame wood building (Virdi, 2013). In this context, in light frame wood structures, the thermal performance of the wall is the key in the energy saving design.

At present, the abundant bamboo resource can be used as an effective substitute for traditional wood in the energy-saving building. Compared with wood, bamboo grows faster (Janssen, 1981), has better dimensional stability (Janssen, 1981), has better mechanical properties (Ahmad, 2005), and its density is similar to that of common wood (Ramage, 2015). Therefore, many kinds of engineering bamboo are developed and applied (Mahdavi, 2011). Xiao (2008) proposed a new type of glued bamboo named Glubam. Its mechanical properties are comparable to wood and raw bamboo. In particular, Li (2013), Xiao (2014) and Wang (2017) have tested the mechanical properties of light frame bamboo shear wall and found that the seismic performance of bamboo shear wall can meet the design requirements for the wood building.

The engineering bamboo material can meet the requirement of engineering application, so it is feasible to use it as a building material. As an alternative component of wood wall, the thermal performance of light frame bamboo wall has not yet been studied very well. Therefore, it is urgent to study the thermal performance of the bamboo wall.

To study the thermal insulation performance of light frame bamboo walls and to promote the application of bamboo as a new material in energy saving construction, the thermal transmittance of two light frame bamboo walls, one light frame wood wall and one light frame bamboo-wood composite wall were studied by using a guarded hot box test. The numerical simulation of steady-state heat transfer of three types of walls was carried out by using the method reported in ISO 6946 and the finite element method. The results highlight the similar performances of wood and bamboo walls and the good capability of the models to predict the experimental results.

2 EXPERIMENTAL TESTS

2.1 *Design and fabrication of the wall models*

Referring to GB 50005-2003 named as Code for Wood Structural Design (GB 50005-2003, 2006), two light frame bamboo walls (sheeting panel in bamboo plywood and framing in Glubam) with different stud thickness, one light frame wood wall (sheeting panel in OSB and framing in SPF) and one light frame hybrid bamboo-wood composite wall (sheeting panel in bamboo plywood and framing in SPF) were studied. According to the requirement of test equipment, the dimensions of the test wall are 1830 × 1830 mm (i.e., high H × wide L).

The tested configurations of the archetype wall in terms of geometrical and material definitions are reported in Figure 1 and Table 1. The wall component is divided into six layers from outdoor to indoor. The first layer is exterior wall finishing composed by the ferro-cement jacket with a thickness of 19 mm. The galvanized U-nail with the distance of 300 mm is used to penetrate the moisture-proof layer and connect with the sheeting panel. The second layer is the air-weather retarder composed by high-density polyethylene fibers (HPF) with a thickness of 0.17 mm. The third layer is the sheeting panel, and the light frame wood wall is realized by 9 mm thick OSB, light frame bamboo-wood composite wall and light frame bamboo wall are all composed by ply-bamboo (PB) with a thickness of 9 mm. The sheeting panel is connected to the frame by common steel nails with a distance of 150 mm. The fourth floor is composed of a frame and rock wool filler, in which rock wool is the thermal insulation material. The frame is composed of studs and top (bottom) plate, in which the light frame wood wall and the light frame bamboo-wood composite wall are all combined with SPF with the size of 38 × 89 mm while two types of Glubam stud thickness (38 × 89 mm and 38 × 140 mm) were used in the light frame bamboo wall. The fifth layer is a vapor retarder realized by polyethylene (PE) with a thickness of 0.2 mm, whose function is to prevent the movement of air and water vapor, and which is bonded to the wall frame by the binder. The sixth layer is the interior wall finish, which is composed of 12 mm thick gypsum board and is connected with the frame by self-tapping screw spacing 150 mm.

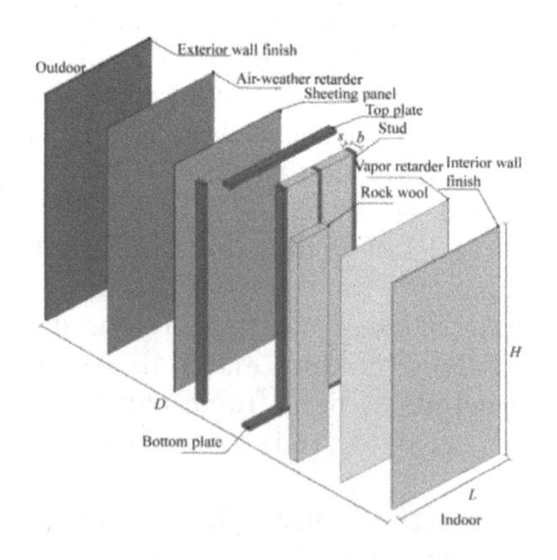

Figure 1. Sketch of the components of the shear wall analyzed.

Table 1. Characteristics of the shear wall analyzed.

		Wall			
Type		A	B		C
b	mm	89	89	89	140
n		5			
s	mm	38			
L	mm	1830			
H	mm	1830			
D	mm	130.37	130.37	130.37	180.37
	1	Ferro-cement 19 mm			
	2	HPF 0.17 mm			
	3	OSB 9 mm PB 9 mm PB 9 mm			
	4a	Rockwool			
	4b	SPF	SPF		Glubam
	5	PE 0.2 mm			
	6	Gypsum board 12 mm			

2.2 *Test scheme and measurement content*

The tests of the thermal insulation performance of the four walls are carried out according to ISO 8990 (1994). The test device is shown in Figure 2. The device consists of four components: cold box, specimen placement frame, hot box and guarded box. The steady-state test is based on one-dimensional heat transfer theory to simulate the heat transfer in walls. The temperature of the guarded box is the same as that of the hot box. The heating power of the guarded box is the heat flowing through the wall. The test temperature is gathered by the thermocouple located on the wall surface in the guarded and cold box (Figure 3). In order to ensure the steady heat transfer of the wall, more than 12 hours measuring period is guaranteed. When the errors of the measured thermal transmittance are less than 1% in two consecutive 3-hour measurement time windows, the test is completed (ISO 8990, 1994). In this test, the temperature of the cold box and the hot/guarded box are set at 0 and 30°C, respectively.

The thermal transmittance (K) is the representative value to show the thermal performance of the energy-saving envelope. Based on a series of experimental data, Formula (1) is used to calculate the thermal transmittance (W/m²K) (ISO 8990, 1994).

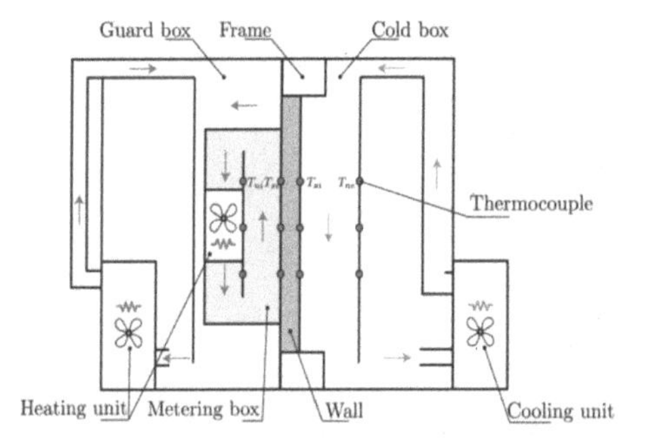

Figure 2. Schematic diagram of the guard hot box testing apparatus.

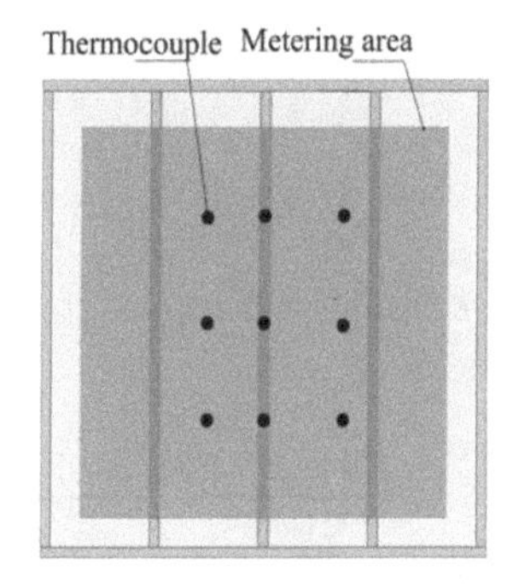

Figure 3. Schematic diagram of the measuring area.

$$K = \frac{\Phi}{A(T_i - T_o)} \quad (1)$$

Where, Φ is the heating power of the guarded box at steady-state condition in W, A is the measuring area in m, T_i and T_o are the temperature of the guarded box and the cold box in °C.

3 TEST AND ANALYSIS

According to the test data in the measurement period in Figure 4 and the average thermal transmittance summarized in Table 2, it can be seen that the thermal transmittance of Wall A89 (A is the type, 89 is the frame thickness, *b*, see Table 1 for definition) is the lowest under the same thickness situation, which indicates that the thermal insulation performance of the Wall A89 is better than that of the Wall B89 and the Wall C89 in which the influence of the sheeting panel in thermal insulation performance is around 2%. In addition, compared with the Wall B89, the thermal transmittance of the Wall C89 is larger because the thermal conductivity of SPF framing is different from Glubam, and the difference of framing materials has around 10% influence in thermal insulation performance. When the stud thickness of the wall C increases from 89 mm to 140 mm, the thermal transmittance decreases by 32%, and the thermal performance improves greatly. In the tested walls, the Wall C140 has the lowest thermal transmittance corresponding to the best thermal performance.

When the light frame bamboo wall is selected as energy-saving building envelope structure in different regions of China, the Wall C89 can be used as building energy-saving wall in hot summer and warm winter areas, mild summer and cold winter areas, and the Wall C140 can be used as energy-saving exterior wall in cold and severe cold areas (GB/T 50361, 2006).

4 SIMULATIONS VS. TESTS

4.1 *ISO 6946 approach*

The ISO 6946 (2007) code provides a method for calculating the average thermal transmittance of the inhomogeneous wall with the following assumptions: two-dimensional thermal

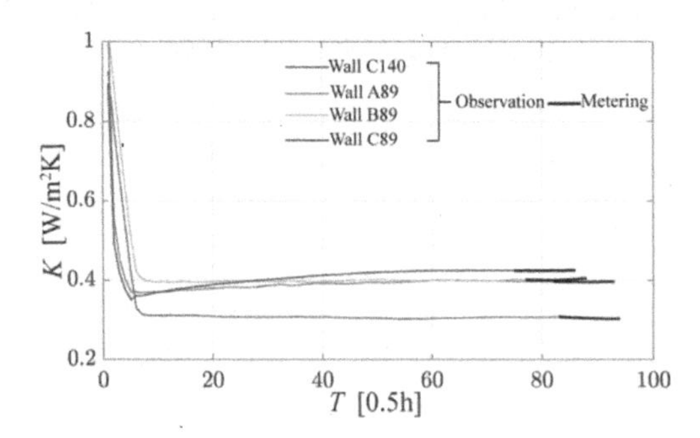

Figure 4. *K* changes with time.

Table 2. *K* of tested wall.

	Wall			
	A89	B89	C89	C140
K [W/m^{2}K]	0.395	0.402	0.457	0.308

analysis and the thermal conductivity of the material isotropic. The formula for calculating the average thermal transmittance is as follows:

$$K = \left[\frac{\left(\sum_m \frac{f_m}{\sum_j R_{m,j} + R_i + R_e}\right)^{-1} + \sum_j \left(\frac{f_m}{\sum_m R_{m,j}}\right)^{-1} + R_i + R_e}{2} \right]^{-1} \tag{2}$$

where m is the part of the wall, j is the layer of the wall, f_m is the percentage of the area of each part in the middle area, $R_{m,j}$ is the thermal resistance of j layer of m part (m²K/W), and $R_{m,j} = d_{m,j}/\lambda_{m,j}$, where $d_{m,j}$ is the material thickness (m), $\lambda_{m,j}$ is the material thermal conductivity (W/mK), R_i and R_e are the internal and external surface heat transfer resistance (m²K/W), respectively.

4.2 *Finite element simulation*

The ANSYS software is used to analyze the steady-state heat transfer in the wall, which follows the first law of thermodynamics. In the steady-state heat transfer process, the heat flow and environment temperature condition do not change with time. In addition, the ANSYS finite element simulation can analyze the three-dimensional thermal analysis of the wall, and the isotropic and anisotropic analysis of the material thermal transmittance can be studied.

The heat balance equation can be expressed as follows:

$$\frac{\partial}{\partial x}\left(K_{xx}\frac{\partial t}{\partial x}\right) + \frac{\partial}{\partial y}\left(K_{yy}\frac{\partial t}{\partial y}\right) + \frac{\partial}{\partial z}\left(K_{zz}\frac{\partial t}{\partial z}\right) + \dddot{q} = 0 \tag{3}$$

where t is the temperature in °C, K_{xx}, K_{yy}, K_{zz} are the thermal conductivity in W/mK, q is the thermal generation rate or internal heat source.

The third boundary condition must be satisfied for the steady-state heat transfer:

Inner surface of the wall :

$$-\lambda \frac{\partial t}{\partial z}\Big|_{z=0} = \alpha_1 (t_i - t_{w2}) \tag{4}$$

External surface of the wall :

$$-\lambda \frac{\partial t}{\partial z}\Big|_{z=\delta} = \alpha_2 (t_{w2} - t_o) \tag{5}$$

where α_1 and α_2 are the convective thermal transmittances of the inner and outer surfaces of the external wall in W/mK, T_i and T_o are the indoor and outdoor temperatures in °C.

In the finite element analysis, the SOLID 70 element is used according to the steady-state thermal analysis and the actual size of the tested wall. The thermal transmittance of ANSYS is calculated as follows:

$$K = \frac{B}{A'(T_i - T_o)} \tag{6}$$

where B is the sum of heat flux in W, A' is the inner (outer) side surface area of the wall in m.

The environmental temperature inside and outside the wall simulated by the finite element method is in agreement with the test environment conditions, which are 30°C and 0°C in hot and cold side respectively. The convection coefficient of inside and outside the wall is 30 W/m²K and 23 W/m²K (ISO 6946, 2007). There is no other temperature constraint around the wall.

4.3 *Test verification and discussion*

In order to verify the reliability of the simulation method, the ISO 6946 and ANSYS finite element method were used to compare with the test results. In both models, the thermal conductivity of wall materials is calculated according to Table 3. The thermal conductivity of the exterior wall finish, air-weather retarder, rock wool, vapor retarder and interior wall finish is isotropic, while the thermal conductivity of wall framing layer and sheeting panel is anisotropic (only for the FEM model while this feature is neglected for the ISO 6946 approach). The thermal conductivity of ferro-cement is in accordance with mortar (Fu, 1997), the thermal conductivity of OSB is in reference Zhou (1989), the thermal conductivity of plybamboo and Glubam is in reference Kiran (2012) and Shah (2016), the thermal conductivity of SPF is in reference Lagüela (2014) and Thunman (2002), the thermal conductivity of rock wool is in reference Abdou (2013), the thermal conductivity of gypsum board is in reference Rahmanian (2009), the thermal conductivity of PE and HPF is in reference Kiran (2012) and Shah (2016), the thermal conductivity of SPF is in reference Lagüela (2014) and Thunman (2002). Looking at Table 3, it can be seen that the thermal conductivity of SPF, OSB, Glubam and plybamboo is larger in the longitudinal direction and smaller in the transversal direction (Wang, 2018).

The thermal transmittances of the four walls shown in Figure 5 are compared with the measured values. The thermal conductivity of the materials is isotropic and anisotropic respectively in the finite element simulation. The calculated results of thermal transmittance are larger than the experimental values, but the difference is smaller.

In the simulation of the thermal transmittances of the light frame wood wall and light frame bamboo-wood composite wall, the result of ISO 6946 is the largest, followed by finite element isotropic and anisotropic models. While in the light frame bamboo walls, the finite element isotropic model has the largest value, followed by ISO 6946 model, and the finite element anisotropic model has the smallest result.

Generally speaking, the finite element model with anisotropic thermal conductivity can predict the thermal transmittance of light frame bamboo/wood wall with more accuracy. Therefore, it is suggested that the finite element model with anisotropic thermal conductivity could be used to analyze the thermal performance of bamboo/wood wall. Besides, a good agreement between the experiments and the models was also found with the better results obtained with the finite element model considering the thermal resistance.

Table 3. Thermal conductivity of material. The number inside the brackets are the thermal conductivity in the transversal direction.

	Wall A	Wall B	Wall C
Sheeting panel	OSB	Playbamboo	Playbamboo
Frame	SPF	SPF	Glubam
Exterior wall finish	0.52		
Air-weather retarder	0.02		
Sheeting panel	0.10(0.17)	0.12(0.20)	0.12(0.20)
Frame	0.09(0.20)	0.09(0.20)	0.15(0.25)
Rock wool	0.037		
Vapor retarder	0.02		
Interior wall finish	0.19		

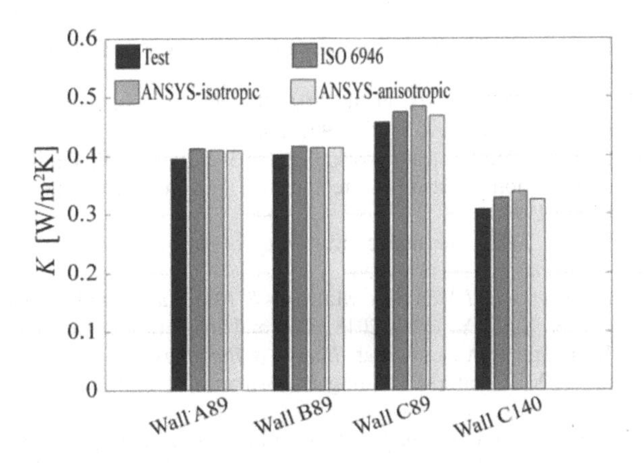

Figure 5. Thermal coefficient of the four types of walls.

5 CONCLUSIONS

The main conclusions that can be made from this study are:

1. The results of finite element analysis and ISO 6946 approach are in good agreement with the experimental results, and the model considering the anisotropy of thermal conductivity of materials can more accurately reflect the thermal insulation performance of the wood or bamboo walls.
2. The thermal performance of the bamboo wall is similar to that of the wood wall and in the thickest configuration (i.e., 140 mm) it can be used as an energy-saving component in all regions in China (Wang, 2018).
3. Given these results in terms of thermal insulation performances, the bamboo walls can be used in energy-saving construction to replace common wood-based walls.

REFERENCES

Abdou, A. & Budaiwi, I. 2013. The variation of thermal conductivity of fibrous insulation materials under different levels of moisture content. *Construction and Building Materials*. 43: 533–544.

Ahmad, M. & Kamke, F.A. 2005. Analysis of Calcutta bamboo for structural composite materials: physical and mechanical properties. *Wood Science and Technology*. 39 (6): 448–459.

Building components and building elements–Thermal resistance and thermal transmittance–Calculation method ISO: 6946-2007. 2007.

Code for Design of Timber Structures: GB 50005-2003. 2006. Beijing. (in Chinese).

Fu, X. & Chung, D.L. 1997. Effects of silica fume, latex, methylcellulose, and carbon fibers on the thermal conductivity and specific heat of cement paste. *Cement & Concrete Research*. 27 (12): 1799–1804.

Janssen, J.J. 1981. *Bamboo in building structures*. Technische Hogeschool Eindhoven.

Kiran, M.C. & Nandanwar, A. & Naidu, M.V, et al. 2012. Effect of density on thermal conductivity of bamboo mat board. *International Journal of Agriculture and Forestry*. 2 (5): 257–261.

Lagüela, S. & Bison, P. & Peron, F, et al. 2014. Thermal conductivity measurements on wood materials with transient plane source technique. *Thermochimica Acta*. 600: 45–51.

Li, Z., Xiao, Y. & Wang, R, et al. 2013. Experimental studies of light-weight woodframe shear walls with ply-bamboo sheathing panels. *Journal of Building Structures*. (09): 142–149 (In Chinese).

Mahdavi, M., Clouston, P.L. & Arwade, S.R. 2011. Development of Laminated Bamboo Lumber: Review of Processing, Performance, and Economical Considerations. *Journal of Materials in Civil Engineering*. 23 (7): 1036–1042.

Rahmanian, I. & Wang, Y. 2009. Thermal Conductivity of Gypsum at High Temperatures – A Combined Experimental and Numerical Approach. *Acta Polytechnica*. 49 (1): 4–6.

Ramage, M., Sharma, B. & Bock, M, et al. 2015. Engineered bamboo: state of the art. *Construction Materials*. 168 (2): 57–67.
Shah, D.U., Bock, M.C.D. & Mulligan, H, et al. 2016. Thermal conductivity of engineered bamboo composites. *Journal of Materials Science*. 51 (6): 2991–3002.
Technical code for partitions with timber framework: GB/T50361-2005. 2006. Beijing. (In Chinese).
Thermal insulation -Determination of steady-state thermal transmission properties - Calibrated and guarded hot box: ISO 8990-1994. 1994.
Thunman, H. & Leckner, B. 2002. Thermal conductivity of wood - models for different stages of combustion. *Biomass & Bioenergy*. 23 (1): 47–54.
Virdi, S. 2013. *Construction Science and Materials*. Materials & Manufacturing Processes. 28 (4): 502–503.
Wang, J.S., Demartino, C. & Xiao, Y, et al. 2018. Thermal insulation performance of bamboo- and wood-based shear walls in light-frame buildings. *Energy & Buildings*. 168.
Wang, R., Xiao, Y. & Li, Z. 2017. Lateral Loading Performance of Lightweight Glubam Shear Walls. *Journal of Structural Engineering*.
Xiao, Y., Li, Z. & Wang, R. 2014. Lateral loading behaviors of lightweight wood-frame shear walls with ply-bamboo sheathing panels. *Journal of Structural Engineering*. 141 (3): B4014004.
Xiao, Y., Shan, B. & Chen, G, et al. 2008. Development of a new type of Glulam-GluBam. Modern bamboo structures.
Zhou, D. 1989. A study of oriented structural board made from hybrid poplar. Physical and mechanical properties of OSB. *European Journal of Wood and Wood Products*. 47 (10): 405–407.

Modern Engineered Bamboo Structures – Xiao, Li & Liu (eds)

 ISBN 978-1-138-35185-1

Deformation properties and microstructure characteristics of bamboo fiber reinforced cement mortar

X. He, J. Yin, Y.Y. Liu, Y. Xiong & N. Lin
Central South University of Forestry and Technology, Changsha, Hunan, China

ABSTRACT: Bamboo fiber strength the effect of deformation performance of cement mortar. In this studied, selected the length of 20 mm bamboo fiber as the research object, and chose different content of bamboo fiber, studied its effects on dry shrinkage performance, discussed pore size distribution characteristics of the bamboo fiber reinforced cement mortar and its effect on macro performance, scanning electron microscope (SEM) and energy spectrum (EPS) used to characterization of the microstructure. The results showed that with the continuous increase content of bamboo fiber, the dry shrinkage rate decrease first and then increase. When the bamboo fiber content increases to 1.5 kg/m^3, its performance improved the best and the number of gel holes was the lowest. The results of SEM and EPS showed that the modified bamboo fiber had a circular cross section and the transition zone between fiber and matrix was compacted.

1 INSTRUCTIONS

Cement-based materials have high compressive strength, but there are disadvantages such as low tensile strength, small ultimate elongation and poor crack resistance. Adding fiber into cement based materials is one of the effective methods to solve the above defects. And the disadvantages of traditional fiber is becoming more and more prominent in its specific application, such as steel fiber to equipment wear and easy corrosion, synthetic fiber prices rising and stability, the high cost and the dispersion problem of carbon fiber, glass fiber threat to human health. Bamboo fiber, as a kind of environmental friendly and renewable resources, compared with the traditional fiber, has widely raw material sources, good processing performance and low energy consumption. Studies have shown that fracture toughness and mechanical properties for mortar improved by adding bamboo fiber. Most of researches focus on the tensile, bending and toughness of composite materials by bamboo fiber, less researches on deformation and microstructure bamboo fiber reinforced cement mortar through the deformation behaviors. The system of fiber reinforced cement based on materials preparation technology has important reference value for the development of new cement-based reinforced materials.

This paper studied the effect of bamboo fiber on the deformation performance and microstructure characteristics of bamboo fiber mortar, explored the pore structure characteristics of bamboo fiber mortar, analyzed the influence of its deformation performance and the mechanism of bamboo fiber mortar based on its microstructure morphology and characteristics.

2 MATERIALS AND METHODS

2.1 *Materials*

Cement: P·O 42.5 grade ordinary Portland cement produced by pingtang cement factory in Changsha, China. Its chemical composition and mechanical properties showed in Table 1 and Table 2.

Fly ash: II grade fly ash provide by power plant in Xiangtan, China.

Sand: Medium sand produced in Xiangjiang River, Hunan. Its gradation curve meets the requirements of Area II.

Water-reducing agent: Polyoxyfuric acid high-performance water-reducing agent, provided by gerilin building materials technology co. LTD in Wuhan, has a water-reducing rate of 30% and a solid content of 20%.

Water: Tap water.

Bamboo fiber: Produced by bamboos industry development co. LTD in Sichuan. The performance parameters are shown in Table 3.

2.2 *Mix ratio design*

Manual carding and pruning were carried out on bamboo fibers, the length of which was 20 mm, and the controlled dosage was 0.5, 1.0, 1.5, 2.0 kg/m^3, respectively. Bamboo fiber soaked by NaOH and dried before adding. Mix sand, cement and bamboo fiber for 2 min, then add water and stir for 2 min. The specific mixing ratio of mortar was shown in Table 4.

2.3 *Drying shrinkage test methodologies*

Refer to Basic performance test of building mortar (JGJ/T70-2009). Firstly, the shrink head was fixed in the hole of both ends of the test mold, the size of test mold was 40 mm × 40 mm × 160 mm, the size of range exposed on the end face of the test piece was 8 ± 1 mm,

Table 1. Chemical compositions of cement.

Chemical Formula	SiO_2	Al_2O_3	Fe_2O_3	CaO	MgO	Na_2O	SO_3
Content	21.38	5.63	3.65	63.72	2.15	1.02	1.75

Table 2. Basic performance parameters of cement.

Flexural strength/MPa		Compression strength/MPa	
3d	28d	3d	28d
5.9	9.4	22.7	50.5

Table 3. Performance parameters of bamboo fiber.

Average Diameter (mm)	Length (mm)	Density (g/cm^3)	Tensile strength (MPa)	Modulus of elasticity (GPa)	Water Absorption (%)
0.18	40~80	1.49	380	35	42

Table 4. The mix proportion of mortar reinforced with bamboo fiber.

Number	Cement (kg/m^3)	Fly ash (kg/m^3)	Bamboo fiber Proportion (kg/m^3)	Sand (kg/m^3)	Water (kg/m^3)	Water reducer (kg/m^3)
A_0	268	114	0	1460	256	3.8
A_1	268	114	0.5	1460	256	3.8
A_2			1.0			
A_3			1.5			
A_4			2.0			

then mixed mortar by cement mortar vibration table vibration close-grained, at 20 ± 5°C indoor, smoothing after 4 h. Mortar maintain 20 ± 2°C, relative humidity of 90% or higher, 7 d after ripping, and numbered, indicate the direction of the test. Put specimens at temperature 20 ± 2°C, relative humidity 60 ± 5%, respectively measure the length of 1, 3, 7, 14, 21, 28, 56 d.

The shrinkage rates in this paper were only calculated for the specimen at each age relative to the length of the specimen at the 1d age. The Equation (1) showed the drying shrinkage rate of mortar.

$$\varepsilon_{\mathrm{at}} = \frac{L_0 - L_t}{L_0} \tag{1}$$

where ε_{at} = the drying shrinkage rate of mortar at day t; L_0 = measured length of mortar sample on day 1, mm; and L_t = the measured length of the mortar sample at day t, mm.

The drying shrinkage rate takes the arithmetic mean value of the measured values of the three samples. When the deviation between one of the three measured values and the mean value was more than 20%, it should be removed. When two values exceed 20% of the average, the results of this group of tests were invalid.

3 RESULTS AND DISCUSSION

3.1 *Dry shrinkage value of bamboo fiber reinforced cement mortar*

Figure 1 showed the dry shrinkage of specimens when the content of bamboo fiber respectively was 0, 0.5, 1.0, 1.5 and 2.0 kg/m^3. In general, the dry shrinkage rate increases greatly at the early stage and gradually flattens out in the later stage. For drying shrinkage rate of growth was larger when curing period before 28 d, then gradually smooth. For cement hydration was inadequate. The content of capillary water and adsorbed water become larger, which lead to rapid speed evaporation. With the growth of curing period, the internal evaporation of water gradually reduce. When kept balance with the intrusion of moisture in the external environment, the dry shrinkage of the mortar become slowly. From Figure 1, compared to specimens with no bamboo fiber, the specimens that dry shrinkage value of decreased by 28.4%, 31.5%, 33.4% and 32.3% respectively, after adding the content of bamboo fiber for 0.5, 1.0, 1.5, 2.0 kg/m^3.

With the increase content of bamboo fiber, the dry shrinkage of mortar decreases first and then increases. For example, when the fiber content increases from 0.5 kg/m^3 to 1.5 kg/m^3, the dry shrinkage value of the specimen continues to decrease, but when the fiber content continues to increase to 2.0 kg/m^3, the dry shrinkage rate of mortar actually increases by 43.8%. For when the length of fiber length was longer, the substrate bonding area larger, and the

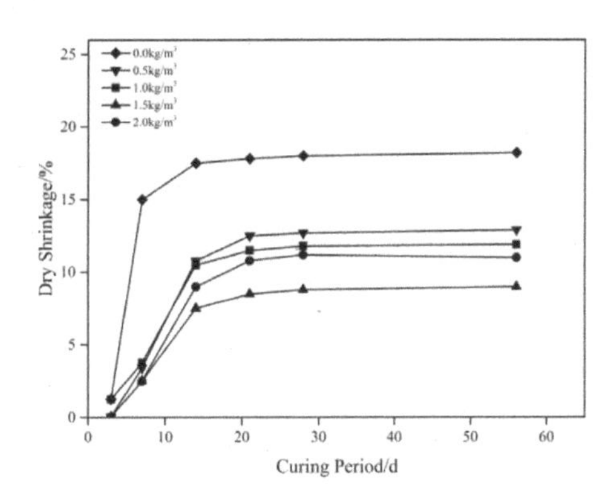

Figure 1. Dry shrinkage value of bamboo fiber reinforced cement mortar.

range of stress transfer became wider. Lower dosage can better limit crack effect. The larger dosage caused negative effect and will make the dry shrinkage rate of the mortar arise.

3.2 *Pore structure distribution characteristics of bamboo fiber reinforced cement mortar*

In order to explain the influence of bamboo fiber content on mortar drying shrinkage from microscopic perspective. Table 5 showed pore structural parameters of mortar. When added bamboo fiber into the mortar, its porosity increased. And with the increased of bamboo fiber content, the porosity will also increase. The addition of bamboo fiber also changes the pore structure inside the mortar matrix. When added bamboo fiber into the mortar, the number of capillary holes that less than 50 nm in the matrix decreased, while the number of harmful and multiharmful holes that the size larger than 50 nm increased. This can explain the strength of specimens decreased when the mortar were mixed with bamboo fiber. When the capillary pore diameter was larger than 50 nm, the additional pressure caused by the surface tension of water was smaller and negligible. However, when the pore diameter was less than 1nm, the water is difficult to lose in the atmosphere, so the shrinkage was mainly caused by the loss of water in the capillary pore with the pore diameter of 1 to 10 nm.

Figure 2 displayed variation of pore size distribution of mortar at curing period of 28d. The number of gel holes with a diameter of 1 to10 nm gradually decreases after adding bamboo fiber into the mortar. For when the bamboo fiber content was 1.5 kg/m^3, the number of gel holes decreases by 28.6% compared with the reference mortar, and the dry shrinkage rate also decreases by 53.7% compared with the reference mortar. The number of gel holes with a diameter of 1to 10nm has an important impact on the shrinkage of the mortar.

The number of gel holes with the pore diameter of 1 to 10 nm increases compared with 1.5 kg/m^3 when the fiber content was 2.0 kg/m^3, resulting in an increase in the dry shrinkage rate. Dry shrinkage was caused by capillary pressure on the pore wall when the evaporable

Table 5. Pore structural parameters of mortar.

Serial Number	Porosity/%	Pore size distribution/%			
		<0.01 μm	0.01-0.05 μm	0.05-1.00 μm	>1.00 μm
A_0	20.34	42.6	35.8	9.2	12.4
A_1	21.85	41.9	33.7	11.9	12.5
A_2	22.05	34.3	32.1	17.3	16.3
A_3	22.14	30.4	31.1	20.1	18.4
A_4	27.21	39.5	38.4	12.7	9.4

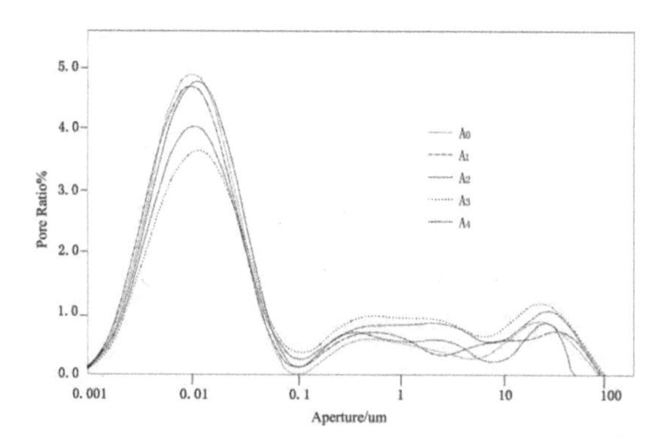

Figure 2. Variation of pore size distribution of mortar at 28d with bamboo fiber content.

water in the capillary hole moves out to the external environment. The change of pore structure can explain the influence of different fiber content on dry shrinkage of mortar.

3.3 *Relationship between pore structure and dry shrinkage of bamboo fiber reinforced cement mortar*

For further discussion for pore size distribution of the bamboo fiber reinforced cement mortar. Fitted the index, such as porosity, average pore diameter and maximum pore diameter, which with dry shrinkage. Figures 3 (a), (b), (c) showed the result of fitting, respectively, porosity, average pore diameter and pore diameter biggest impact on the rate of drying shrinkage, the fitting results were 0.97, 0.99, 0.98, showed that pore structure for drying shrinkage has important effect.

3.4 *Microstructure characterization of bamboo fiber reinforced cement mortar*

Figures 4 showed the SEM pictures of the untreated bamboo fiber and the bamboo fiber treated with alkali. As can be seen from Figure 4 (a), there were more colloids on the surface of untreated bamboo fibers, and the small and medium fiber bundles of bamboo fibers were bound together by colloids to ensure own strength and toughness. Figure 4 (b) showed that each small fiber bundle of untreated bamboo fibers was tightly bonded and the fiber integrity was relatively good. When after alkali treatment, hemicellulose and lignin in bamboo fiber were dissolved, and the connection between fiber bundles was weakened, which will lead to the decrease of tensile strength of bamboo fiber itself. Figure 4 (c) displayed the gum on the surface of bamboo fiber significantly decreased after alkali treatment, and some small fiber bundles were peeled off. It can be clearly seen from Figure 4 (d) that each independent fiber bundle with a circular cross section in the fiber.

Figures 5 showed the bamboo fiber mortar and microscopic pictures, respectively. It can be seen from Figure 5 (a) that the bamboo fiber pulled out of the mortar matrix remains intact,

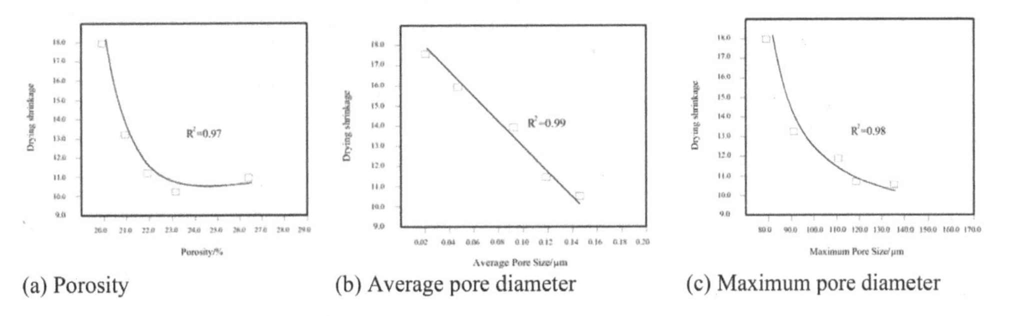

(a) Porosity (b) Average pore diameter (c) Maximum pore diameter

Figure 3. The relationship between pore structure characteristic parameters and dry shrinkage of bamboo fiber reinforced cement mortar.

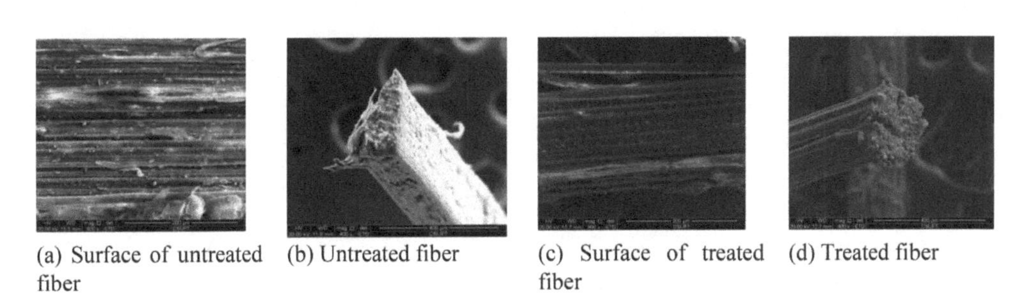

(a) Surface of untreated fiber (b) Untreated fiber (c) Surface of treated fiber (d) Treated fiber

Figure 4. Microstructure of bamboo fiber.

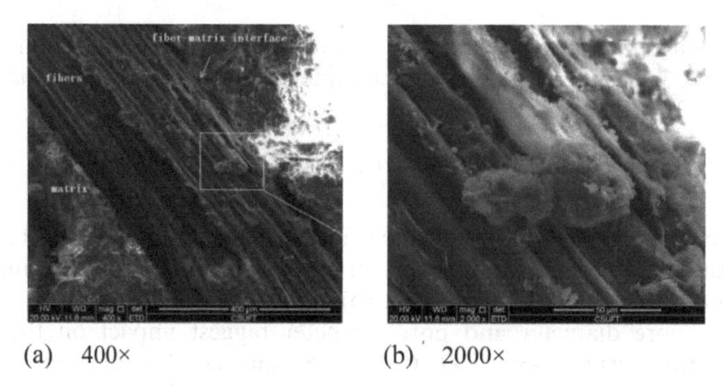

(a) 400× (b) 2000×

Figure 5. Microstructure of mortar reinforced with bamboo fiber.

and there is no obvious sign of corrosion except a small amount of hydration product deposition on the surface. EDS energy spectrum analysis is conducted on the dotting point 1 in Figure 5 (b), and the results were shown in Table 6 and Figure 6 respectively. The chemical composition of point 1 is mainly carbon and oxygen, followed by calcium, silicon and aluminum. For the plant fiber is composed of organic macromolecular compounds, so the chemical composition of point 1 on the fiber surface contains high content of carbon and oxygen, while the hydration products on the fiber surface were less, mainly calcium hydroxide crystals, calcium silicate hydrated and calcium aluminate hydrated gel.

Figure 5 (b) showed that more hydration products were attached to the fiber surface, which compacts the interface transition zone between fiber and matrix. At this time, except for slight signs of mineralization, the fiber almost maintains its initial surface shape. The interface area between the fiber and the matrix was dense, but at the same time, the fiber structure was also damaged to a certain extent. The cavities between the fiber bundles were filled with thick slate-like hydration products. When bamboo fiber mortar was immersed in water, free ions dissolved from the matrix infiltrate into the fiber cavity, forming ettringite and calcium hydroxide crystals, which mineralize the fiber and increase its brittleness.

Table 6. Elementary compositions of each observation point at EDS test.

Elementary Composition/%				
C	O	Al	Si	Ca
27.69	55.96	2.57	4.42	6.43

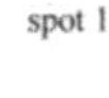

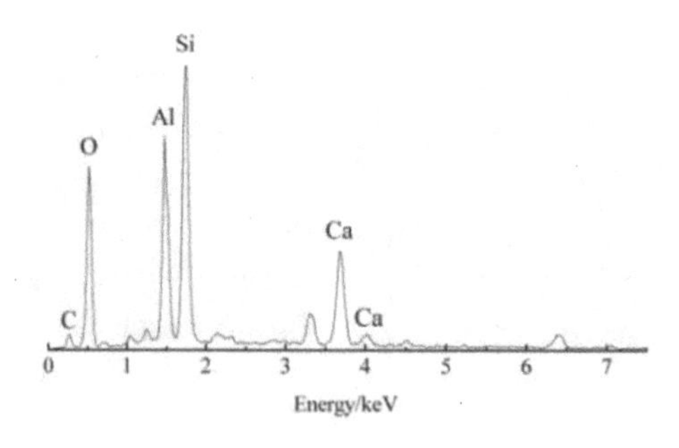

Figure 6. EDS patterns.

4 CONCLUSION

(1) With the continuous increase of bamboo fiber content, the dry shrinkage performance and cracking area of cement mortar decreased first and then increased, and the crack reduction coefficient increased first and then decreased. When the bamboo fiber content reached 1.5 kg/m^3, the dry shrinkage rate was the minimum for 7.80%.
(2) With the continuous increase of bamboo fiber content, the number of gel holes of cement mortar decreases first and then increases. When the bamboo fiber content reached 1.5 kg/m^3, the number of gel holes was the least. The fitting results show that the porosity, maximum pore diameter and average pore diameter were correlated with the dry shrinkage of cement mortar.
(3) The SEM and EPS results showed that after alkali treatment, hemicellulose and lignin in bamboo fiber was dissolved. And the connection between the fiber bundles were weakened, glue on the surface of the bamboo fiber decreased significantly, smaller fiber bundle, stripped away, fiber in separate for circular cross section of fiber bundles, dense fibers and cement mortar matrix interface area.

ACKNOWLEDGMENTS

This work was supported by the National key R & D program (No. 2016YFC0700801-01).

REFERENCES

Angel, M. & Lopez, B. 2013. Surface treated polypropylene (PP) fibers for reinforced concrete. *Cement and Concrete Research* 54: 29–35.

Azwa, Z.N. & Yousif, B.F. 2013. A review on the degradability of polymeric composites based on natural fibres. *Materials and Design* 47: 424–442.

Chakraborty, S. & Kundu, S.P. 2013. Improvement of the mechanical properties of jute fibre reinforced cement mortar: A statistical approach. *Construction and Building Materials* 38: 776–784.

Giuseppe, T. & Fausto, M. 2014. Cracking beahvior in reinforced concrete members with steel fibers: A comprehensive experimental study. *Cement and Concrete Research* 68 (1): 24–34.

Wang, J.P. & Meyer, C. 2014. Improving degradation resistance of sisal fiber in concrete through fiber surface treatment. *Applied Surface Science* 289: 511–523.

Jodilson, A.C. & Paulo, R.L.L. 2014. Compressive stress-strain behavior of steel fiber reinforced-recycled aggregate concrete. *Cement and Concrete Composites* 46: 65–72.

Kim, H. & Okubo, K. 2013. Influence of fiber extraction and surface modification on mechanical properties of green composites with bamboo fiber. *Journal of Adhesion Science and Technology* 27 (12): 1348–1358.

Liu, D.G. & Song, J.W. 2012. Bamboo Fiber and Its Reinforced Composites: Structure and Properties, *Cellulose* (19): 1449–1480.

Sedan, D. & Pagnoux, C. 2008. Mechanical properties of hemp fibre reinforced cement: Influence of the fibre/matrix interaction, *Journal of the European Ceramic Society* 28 (1): 183–192.

Viviane, C.C. & Sergio, F.S. 2014. Potential of bamboo organosolv pulp as a reinforcing element in fiber-cement materials. *Construction and Building Materials* 72: 65–71.

Modern Engineered Bamboo Structures – Xiao, Li & Liu (eds)
© 2020 Taylor & Francis Group, London, ISBN 978-1-138-35185-1

Investigation on properties of bamboo fiber reinforced cement mortar

J. Yin, Y.Y. Liu & J.W. Yang
Central South University of Forestry and Technology, Changsha, Hunan, China

ABSTRACT: Adding fibers to cement-based materials is one of the effective ways to improve their brittleness and shrinkage cracking. In this study, bamboo fiber mortar was taken as the research object. Firstly, the mixing technology of bamboo fiber reinforced cement mortar was optimized, and then the influence of bamboo fiber content and length on the mechanical properties and plastic shrinkage of mortar was discussed. Finally, the durability of bamboo fiber reinforced cement mortar with good mechanical properties was studied. The results show that the content and length of bamboo fiber have an important influence on the rupture strength and compressive strength of mortar. Based on the strength, the best effect is obtained when the content and length of bamboo fiber are 1.5 kg/m^3 and 10 mm respectively. Bamboo fiber can improve the plastic shrinkage of the mortar and reduce its cracking area. The rupture strength and tenacity of bamboo fiber mortar will gradually decrease after the dry-wet circles. After alkali treatment, the durability of bamboo fibers will be improved to some extent.

1 INTRODUCTION

Cement-based materials have high compressive strength, but also have some shortcomings, such as low tensile strength, low ultimate elongation and poor crack resistance. With the increase of compressive strength, the shrinkage and brittleness of cement-based materials become more prominent. In order to solve these problems, the above defects are usually improved by incorporating fibers into them. As an environmentally friendly and renewable resource, bamboo fiber has the advantages of wide source of raw materials, fast regeneration, good processing performance and low energy consumption compared with traditional reinforced fiber material. Pakotiprapha was the first to study the flexural strength of naturally cured bamboo fiber reinforced cement. Coutts et al. compared the influence of fiber weight content on the flexural strength and fracture toughness of plates. The results showed that when the fiber content was 10%, the flexural strength could reach 20 MPa, but the fracture toughness was relatively low. It has been reported that the addition of bamboo fiber can significantly improve the tensile and bending properties of the composite. However, some studies have pointed out that although the early compressive and flexural strength of cement-based materials can be improved by adding bamboo fibers, the later strength of bamboo fibers tends to decrease due to the corrosion of bamboo fibers in cement-based alkaline media. In addition to improving the tensile and bending strength of the cement matrix, bamboo fiber can also significantly improve the brittle failure of cement-based materials. The research results of can support this conclusion. In order to offset the adverse effect of bamboo fiber on cement hydration, some scholars modified bamboo fiber with NaOH solution to improve its dispersibility, wettability and adhesion with cement.

2 OBJECTIVE

Based on the optimized preparation technology of bamboo fiber mortar, the effects of bamboo fiber length and content on the mechanical properties and deformation properties of bamboo fiber were studied, and the deterioration law of bamboo fiber mortar under wet-dry cycle was also explored, and the improvement measures of durability of bamboo fiber mortar were put forward.

3 EXPERIMENTAL PROGRAM

3.1 *Material*

Cement: P·O 42.5 grade ordinary Portland cement produced by Pingtang Cement Plant, Changsha, Hunan Province. Its chemical composition and mechanical properties are shown in Tables 1-2.

Fly ash: Class II fly ash provided by Hunan Xiangtan Power Plant.

Sand: Medium sand produced in Xiangjiang River, Hunan. Its gradation curve meets the requirements of Area II.

Water reducer: Polysomic acid high-performance water reducer, provided by Wuhan Greylin Building Materials Technology Co., Ltd., water reduction rate is 30%, solid content is 20%.

Water: tap water.

Bamboo fiber: Produced by Sichuan bamboos industry development co., LTD. The performance parameters are shown in Table 3.

3.2 *Experimental method*

3.2.1 *Design of mix content*

For the fiber reinforced cement mortar, the uniform dispersion of fiber in the mortar matrix is the key to guarantee the workability, mechanical properties and durability of the mortar. In order to determine the optimum mixing process of bamboo fiber mortar, three kinds of mixing processes were used to prepare specimens, which were recorded as S1, S2 and S3, respectively. Among them, in method S1, sand and bamboo fibers were added to mix for 2 minutes, then cement was added to mix for 1 minute, and finally water was added to mix for 1 minute. In method S2, sand and cement were added to mix for 1 minutes, and then added suspension made up of bamboo fiber, water reducer and water which was prepared 25 min

Table 1. Chemical compositions of cement.

Chemical formula	SiO_2	Al_2O_3	Fe_2O_3	CaO	MgO	Na_2O	SO_3
Content (%)	21.38	5.63	3.65	63.72	2.15	1.02	1.75

Table 2. Basic performance parameters of cement.

Flexural strength/MPa		compression strength/MPa	
3d	28d	3d	28d
5.9	9.4	22.7	50.5

Table 3. Performance parameters of bamboo fiber.

Average diameter (mm)	Length (mm)	Density (g/cm^3)	Tensile strength (MPa)	Modulus of elasticity (GPa)	Water Absorption (%)
0.18	40~80	1.49	380	35	42

before the test and mixed for 3min. In method S3, sand, cement and bamboo fibers were added to mix for 2 minutes, then water was added to mix for 2 minute. Researches have shown that the length of fiber added to the mortar is generally 3~19 mm, and the content is generally 0.9~1.2 kg/m^3. Therefore, bamboo fiber was artificially combed and trimmed, and the length of bamboo fiber was 5, 10, 15, and 20 mm, respectively, as shown in Figures 1~3. The content was controlled as 0.5, 1.0, 1.5 and 2.0 kg/m^3, respectively. The specific mix ratio of mortar is shown in Table 4. The number before Si (i=1, 2, 3) in the specimen number is the fiber content, and the number after is the fiber length. For example, 1.0 S1-10 means that the fiber content is 1.0 kg/m^3, the fiber length is 10 mm, and the mortar sample is prepared by S1 stirring process.

3.2.2 *Sinking degree and apparent density test*

The test was carried out in accordance with the consistency test and the apparent density test method in "Standards for Testing Basic Performance of Building Mortars" (JGJ/T70-2009).

3.2.3 *Mechanical property test*

1. Compression strength test
   The compressive strength test was carried out by YA - 300A microcomputer controlled automatic pressure testing machine manufactured by Wuxi Xinluda Instruments and Equipment Co., Ltd. The loading rate was 2400 ± 200 N/S.
2. Flexural strength test
   Refer to "Cement mortar strength test method (ISO method)" (GB/T17671-1999 idt IS0679: 1989), the flexural strength test is carried out by KZJ-5000 cement electric bending test machine produced by Wuxi Xidong Building Materials Equipment Factory. The loading rate was 50 ± 10 N/S and the specimen size was 40 mm × 40 mm × 160 mm.

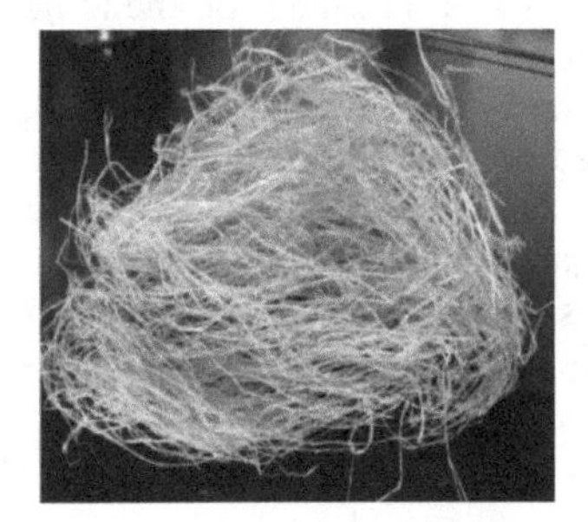

Figure 1. Natural bamboo fiber.

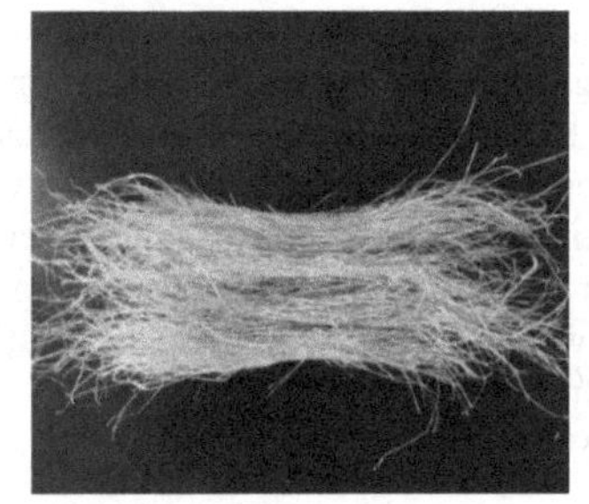

Figure 2. Combed bamboo fiber.

Figure 3. Snipped bamboo fiber.

Table 4. The mix content of mortar reinforced with bamboo fiber.

Number	Cement (kg/m^3)	Flyash (kg/m^3)	Bamboo fiber Length (mm)	Bamboo fiber Content (kg/m^3)	Sand (kg/m^3)	Water (kg/m^3)	Water reducer (kg/m^3)
S	268	114	——	——	1460	256	3.8
1.0S1-10			10	1.0			
1.0S2-10	268	114	10	1.0	1460	256	3.8
1.0S3-10			10	1.0			
0.5S3-10	268	114	10	0.5	1460	256	3.8
1.5S3-10			10	1.5			
2.0S3-10			10	2.0			
1.5S3-5			5	1.5			
1.5S3-15	268	114	15	1.5	1460	256	3.8
1.5S3-20			20	1.5			

3.2.4 *Deformation performance test*

Refer to "Technical Specification for Fiber Reinforced Concrete Structures" (CECS 38: 2004), 600 mm × 600 mm × 20 mm wood mould was used with plastic film lining at the bottom of the mould to prevent water from flowing away from the bottom. After pouring the mortar mixture into the wood mould, cover it with plastic film and place it in the room at 20°C for 2 hours, then remove the film, each specimen is blown by an electric fan, the wind speed is 0.5 m/s. The ambient temperature of the test is 20 ± 2°C and the relative humidity is less than 60%.

Cracks should be visible to the naked eye, and their length should be measured with a steel ruler (for curvilinear cracks, fine lines can be used to simulate their trajectories, and then the length of fine lines can be measured as the length of cracks). Three large-width values were read in the direction of the crack with a reading microscope, and the average value was taken as the nominal maximum width of the crack. The total fracture area and fracture reduction coefficient are calculated as follows:

$$A_{cr} = \sum_{i=1}^{n} w_{i,\max} l_i \tag{1}$$

$$\eta = \frac{A_{mcr} - A_{fcr}}{A_{mcr}} \tag{2}$$

Where A_{cr}: the nominal total area of the crack, the fiber mortar specimen is recorded as A_{fcr}, and the base mortar specimen is recorded as A_{mcr}, mm^2; $w_{i,max}$: the nominal maximum width of the crack i; l_i: The length of the crack i, mm.

3.2.5 *Dry-wet cycle test of bamboo fiber mortar*

The bamboo fiber was surface treated with NaOH. The effect of alkali treatment on the long-term durability of bamboo fiber mortar was studied by dry-wet cycle test, taking the flexural strength and compressive strength of samples before and after the cycle as evaluation indexes. The experiment was conducted by referring to the dry-wet cycle test method proposed by Brazilian scholar Lima Jr H C. The specific steps were as follows: immersing in 20°C water for 20 h, drying in room temperature air for 4 h, drying in an oven at 100°C for 20 h, and cooling at room temperature for 4 h for one cycle. The strength changes of bamboo fiber mortar after 0, 7, 15, 30, 45, 60 and 90 dry-wet cycles were tested to investigate the performance deterioration of bamboo fiber mortar under dry-wet cycles.

4 RESULTS AND DISCUSSION

4.1 *Effect of different mixing processes on the properties of bamboo fiber reinforced mortar*

The properties of bamboo fiber mortar mixtures and the strength of hardened mortar were vary with different mixing processes. The experimental results are shown in Table 5.

As can be seen from Table 5, for each of the above stirring processes, the addition of bamboo fibers reduced the sinking degree of mortar, which means that the fluidity also decreased. This is caused by the water absorption of bamboo fibers, which reduced the free

Table 5. Performance of mortar reinforced with bamboo fiber under the different mixing process.

Mixing method	Sinking degree (mm)	Apparent density (kg/m^3)	28d flexural strength (MPa)	28d compressive strength (MPa)
S	92	2100	5.5	24.1
1.0S1-10	60	2039	4.5	16.6
1.0S2-10	65	2046	4.6	19.0
1.0S3-10	73	2059	5.2	22.3

water in mortar. Under the conditions of the same amount and length of bamboo fiber, the sinking degree of mortar under S3 mixing process was larger than that of S1 and S2. This shows that the bamboo fibers could be more evenly separated in mortar by using S3 stirring method, while under the stirring conditions of S1 and S2, the bamboo fibers were unevenly distributed, entangled and aggregated, and the introduction of more porous spaces reduced the apparent density of mortar, thus affecting the flexural and compressive properties of hardened mortar. Therefore, the optimized stirring process of S3 can ensure that bamboo fibers are evenly dispersed in the matrix, and at the same time, the flow performance and mechanical properties of bamboo fiber mortar can reach the best.

4.2 *Effect of different bamboo fiber length and content on mechanical properties of mortar*

4.2.1 *The influence of bamboo fiber content*

The mortar specimens were formed by the optimized mixing process according to the mixing ratio in Table 4, and their mechanical properties were tested. The effect of fiber content on the strength of mortar was shown in Figure 4.

From Figure 4, it can be seen that after adding bamboo fiber into the mortar, its flexural strength and compressive strength both decrease compared with the reference mortar, but the degree of decrease varies with the fiber content. When the fiber content is 0.5 kg/m^3, the flexural strength and compressive strength were reduced by 8% and 11%, respectively, compared with the reference mortar. According to the theory of fiber spacing, the more the number of fibers per unit area, the more obvious the anti-cracking and reinforcing effect of the fibers. Therefore, when the fiber content increased from 0.5 kg/m^3 to 1.5 kg/m^3, the flexural and compressive strength of mortar increased with the increase of the fiber content. However, when the fiber content exceeds 1.5 kg/m^3, the flexural and compressive strength of mortar decreased with the increase of the fiber content. On the one hand, it is due to the excessive number of fibers per unit area, the obvious fiber agglomeration phenomenon, and the introduction of more pores and defects. On the other hand, due to the continuous increase of fiber content, there is not enough cementitious material in the system to wrap it, which affects the bonding strength of fiber and matrix.

4.2.2 *The influence of bamboo fiber length*

The influence of fiber length on the mechanical properties of mortar is shown in Figure 5.

As illustrated in the figure, when the fiber content was the same and the fiber length was 5~10 mm, the flexural strength and compressive strength of the mortar increased with the increase of the fiber length. When the fiber length was 10 mm, the flexural and compressive strength of the mortar reaches the maximum, which were 5.4 MPa and 22.6 MPa, respectively.

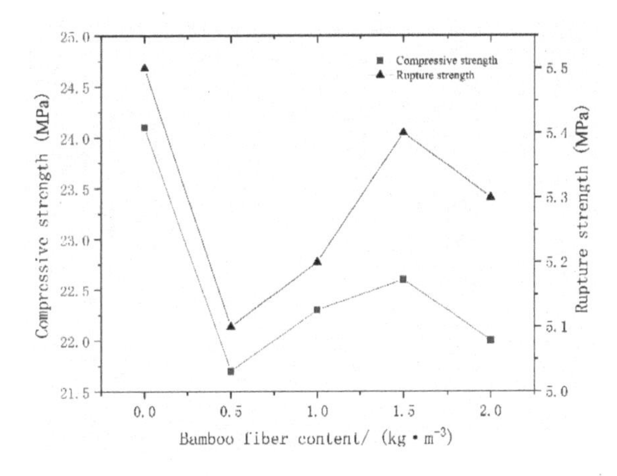

Figure 4. Effect of fiber content on mortar strength.

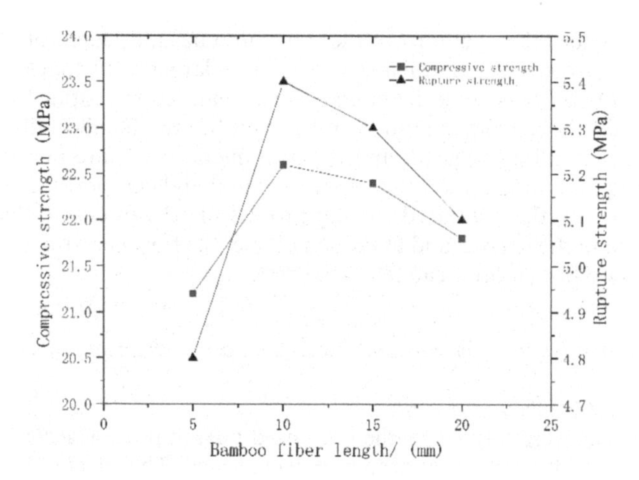

Figure 5. Effect of fiber length on mortar strength.

When the fiber length was 10~20 mm, the flexural strength and compressive strength of mortar decreased with the increase of fiber length. Fiber plays a "bridge" role in mortar. When micro-cracks occur in the mortar under load, the matrix transfers stress to the fiber. Fibers are pulled out or broken from the matrix by debonding with the matrix, which consumes a large amount of energy, thus improving the flexural strength of mortar. When the length of fiber is short, the effect of fiber on load transfer is not obvious, and the contribution to the strength of fiber mortar is not significant. However, when the fiber length exceeds the critical fiber length, the probability of fiber agglomeration increases greatly, and the fiber tangles in the matrix, resulting in an increase in void fraction and a decrease in strength. Therefore, the length of fibers should not be too short or too long.

4.3 *Effect of different bamboo fiber length and content on deformation properties of mortar*

4.3.1 *The influence of bamboo fiber content*

The content of bamboo fiber has a significant effect on the plastic shrinkage cracking of mortar. Figure 6 shows that with the increase of fiber content, the crack area of matrix decreased gradually, and the crack reduction coefficient increased gradually. This is because the plastic shrinkage cracking of mortar is caused by the fact that the capillary shrinkage stress caused by water loss is greater than its plastic tensile strength. When fibers are added

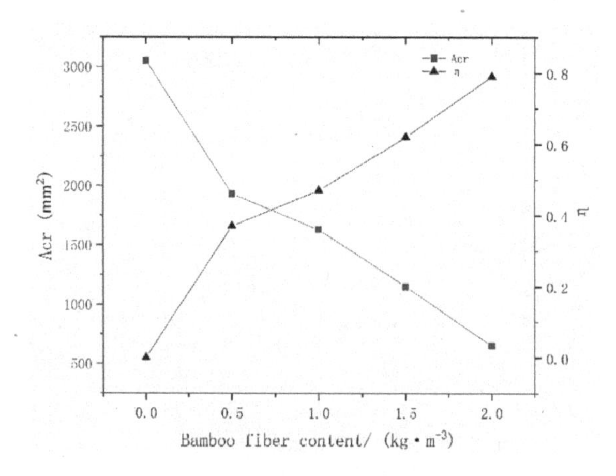

Figure 6. Effect of fiber content on plastic shrinkage of mortar.

into the matrix, on the one hand, the plastic tensile strength of the matrix can be improved by the interfacial cohesive force and mechanical interaction between the fiber and the matrix, on the other hand, depending on the water absorption of bamboo fiber, the stored water can supplement the water lost in capillary, and reduce the shrinkage stress caused by water loss in capillary. Under the combined action of the two factors, the plastic shrinkage cracking of mortar decreases or even disappears.

4.3.2 *The influence of bamboo fiber length*

From Figure 7, it can be seen that with the increase of fiber length, the plastic shrinkage cracking area of mortar decreased gradually, and the crack reduction coefficient increased gradually. It is worth noting that when the fiber length was increased from 5 mm to 15 mm, the area of plastic shrinkage cracking of the mortar was reduced by 61%, and when the length was continuously increased from 15 mm to 20 mm, the cracking area was only reduced by 45%. This indicates that the fiber length has a great influence on the plastic shrinkage of mortar. When the fiber length was less than 15 mm, the increase of fiber length had a great effect on improving the plastic crack resistance of mortar. When the length of the fibers was longer than 15 mm, its effect on improving the plastic crack resistance of the matrix decreased gradually. The reason for the above phenomenon is that, under certain conditions, increasing the fiber length can improve the bonding area between the fiber and the mortar matrix and improve the plastic tensile strength of the matrix. Additionally, the longer fibers have a better ability to cross cracks and bear tension. Compared with short fibers, it takes more energy to bypass long fibers when cracks extend. Therefore, the plastic crack resistance of matrix is improved. This effect is more obvious in the short term of fiber length, which is consistent with the conclusion of literature. If the fiber length is too long, it tends to agglomerate, and more defects are introduced, thus reducing its effect on improving the plastic crack resistance of the matrix.

4.4 *Study on the durability of bamboo fiber reinforced mortar*

According to previous studies, the mechanical properties of bamboo fiber mortar are better when the length of fiber is 10 mm. Therefore, the length of bamboo fiber used in durability test is 10 mm.

4.4.1 *Compressive strength*

The compressive strength of bamboo fiber mortar under different bamboo fiber content increased with the number of cycle times, which was related to the hydration degree of the matrix. As shown in Figure 8, taking the bamboo fiber content of 1.5 kg/m^3 as an example,

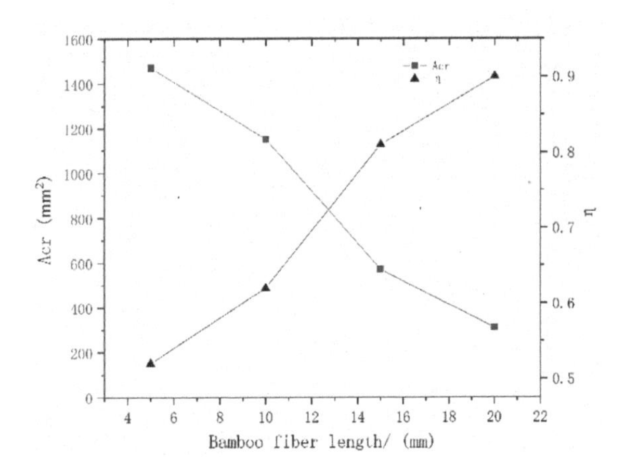

Figure 7. Effect of fiber length on plastic shrinkage of mortar.

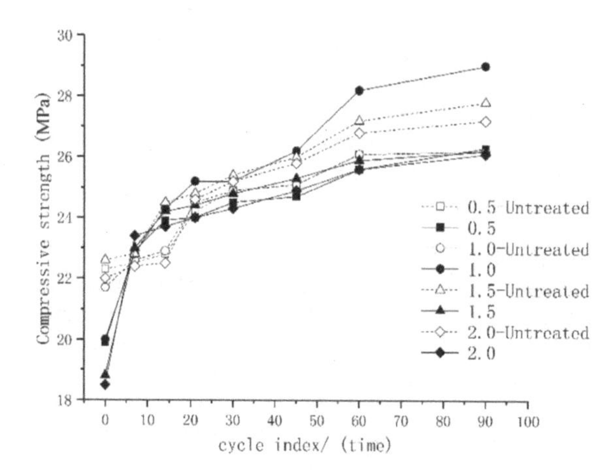

Figure 8. The change of compressive strength of bamboo fiber mortar after dry-wet circles.

the compressive strength of the untreated and alkali-treated bamboo fiber mortar gradually increased with the increase of the of cycle times, and the increase range of 0-14 cycles is larger than that of other cases. After 14 cycles, the compressive strength of alkali-treated bamboo fiber reinforced mortar and untreated bamboo fiber reinforced mortar increased by 28.7% and 8.4% respectively compared with the reference group (the number of cycles was 0). However, when the cycle times exceeded 14, the increase of compressive strength of mortar gradually slowed down. Besides, under the same cycle times, the compressive strength of untreated bamboo fiber mortar was slightly higher than that of alkali-treated bamboo fiber mortar. This is because with the increase of cycle times, the cement hydration is more and more full, the matrix is more and more dense and the compressive strength is also gradually increased. While the compressive strength of alkali-treated bamboo fiber mortar is slightly lower than that of untreated bamboo fiber mortar due to the loss of the strength of the fiber itself.

4.4.2 *Flexural strength*

As it can be found from Figure 9 that the change trend of the flexural strength of mortar with the increase of dry- wet cycles was approximately the same under different bamboo fiber content. Taking the bamboo fiber content of 1.5 kg/m^3 as an example, the flexural strength of bamboo fiber mortar increased first and then decreased with the increase of dry-wet cycles, regardless of whether the bamboo fiber surface was treated or not. When the number of cycles

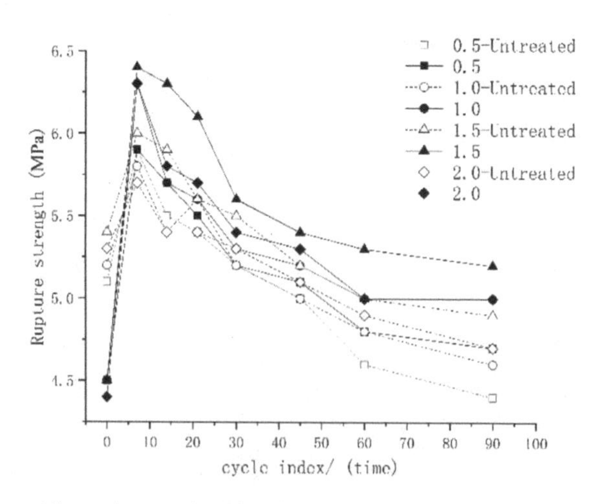

Figure 9. The change of flexural strength of bamboo fiber mortar after dry-wet circles.

was less than 7, the flexural strength of the mortar was higher than that of the reference group. When the number of cycles was 7, the flexural strength of the mortar reached the maximum. At this time, the flexural strength of alkali-treated bamboo fiber mortar and untreated bamboo fiber mortar was 42.2% and 11.1% higher than that of the reference group, respectively. When the number of cycles was greater than 7, the flexural strength of the bamboo fiber mortar decreased gradually with the increase of cycles, but under the same cycle number, the flexural strength of the bamboo fiber mortar treated with alkali was greater than that of the untreated bamboo fiber mortar. After 90 cycles, the flexural strength of untreated bamboo fiber mortar decreased to 90.7% of the reference group, while the flexural strength of alkali-treated bamboo fiber mortar was still 15.5% higher than that of the reference group. In addition, it can also be known that, regardless of the amount of bamboo fiber, when the number of cycles was 0, the flexural strength of alkali-treated bamboo fiber mortar was lower than that of untreated bamboo fiber mortar. This is due to the insufficient hydration of the matrix before the dry-wet cycle test and the loss of the strength of bamboo fiber after alkali treatment. With the increase of cycle times, the colloid such as hemicellulose and lignin on bamboo fiber surface were removed after alkali treatment, so that the hydration products of cement can be arranged more closely on the surface of bamboo fibers and the thick crystalline layer is formed on the surface of the bamboo fiber, leads to that it can better bond with the matrix. Therefore, the long-term performance of the bamboo fiber mortar is better than that of the untreated bamboo fiber mortar.

5 CONCLUSIONS

(1) The optimum mixing process to ensure the dispersibility of bamboo fibers in mortar was selected. That is, firstly, sand, cement and bamboo fibers are pre-mixed for 2 mins, then add water and stir for 2 mins, which is conducive to the uniform dispersion of bamboo fiber in the matrix and improves the working and mechanical properties of bamboo fiber mortar;

(2) The content and length of bamboo fiber have an important impact on the flexural strength and compressive strength of bamboo fiber reinforced mortar. Based on the strength consideration, the best effect is achieved when the content and length of bamboo fiber are 1.5 kg/m^3 and 10 mm respectively;

(3) Bamboo fiber can significantly reduce the plastic cracking degree and crack width of mortar. Under certain conditions, increasing the fiber content and length can improve the plastic tensile strength of mortar;

(4) The flexural strength and toughness of bamboo fiber mortar will decrease gradually after dry-wet cycling. The alkali treatment of bamboo fiber can improve the durability of bamboo fiber mortar to a certain extent.

ACKNOWLEDGMENTS

This work was supported by the National key R & D program (No. 2016YFC0700801-01).

REFERENCES

Bi, J.F. & Wu, H.Q. 2014. Study on mechanical properties of bamboo fiber reinforced mineral Powder-Fly ash cement-based materials. *Journal of Guangxi University of Science and Technology* 25(2): 26–30.

Chakraborty, S. & Kundu, S.P. 2013. Improvement of the mechanical properties of jute fiber reinforced cement mortar: A statistical approach. *Construction and Building Materials* 38: 776–784.

Coutts, R.S.P. & Ni, Y. 1994. Air-cured bamboo pulp reinforced cement. *Journal of Materials Science Letters* 13(4): 283–285.

Coutts, R.S.P. & Ni, Y. 1995. Autoclaved Bamboo Pulp Fibre Reinforced Cement. *Cement and Concrete Composites* 17(2): 99–106.

Dawood, E.T. & Ramli, M. 2010. Development of high strength flowable mortar with hybrid fiber. *Construction and Building Materials* 24(6): 1043–1050.

Li, G.Z. & Tian, Y. 2009. Flexural strength and plastic shrinkage cracking resistance of aramid fiber mortar. *Journal of Building Materials* 12(1): 93–95.

Li, M. & Liu, M. 2014. Study on toughening and cementing well cement stone by alkali treatment modified bamboo fiber. *Functional materials* 45(13): 13087–13091.

Lima, Jr. H.C. & Willrich, F. L. 2008. Durability analysis of bamboo as concrete reinforcement. *Materials and Structures* 41(5): 981–989.

Liu, Y. & Zhang, X.L. 2012. Experimental study on mechanical properties and durability of bamboo fiber/cement composites. *Journal of Henan University of Technology* (Natural Science Edition) 31(1): 104–108-98.

Ma, Y.P. & Zhu, B.R. 2002. Effect of fiber parameters on plastic shrinkage and cracking properties of cement mortar. *Journal of Building Materials* 5(3): 220–224.

Ma, Y.P. & Zhu B.R. 2003. The relationship between plastic tensile strength and shrinkage cracking of cement mortar. 2003. *Journal of Building Materials* 6(1): 20–24.

PakotiPrapha, B. & Pama, R.P. 1983. Behaviour of a bamboo fibre cement paste composite. *Journal of Ferrocement* (13): 235.

Yao, W. & Li, Z.J. 2003. Flexural behavior of bamboo-fiber-reinforced mortar laminates. *Cement and Concrete Research* 33(1): 15–19.

Ye, Y.W. & Xian, D.G. 1998. Bamboo Fiber and Coconut Fiber Reinforced Cement Composites. *Journal of Composite Materials* 15(3): 92–98.

Components & Connections

Modern Engineered Bamboo Structures – Xiao, Li & Liu (eds)
© 2020 Taylor & Francis Group, London, ISBN 978-1-138-35185-1

Effect of opening on OSB webbed bamboo I-shaped beam

N.H. Zheng, G. Chen, B. He, Z. Yuan & H. Zhao
Nanjing Forestry University, Nanjing, Jiangsu, China
National Engineering Research Center of Biomaterials, Nanjing Forestry University, Nanjing, Jiangsu, China

ABSTRACT: Laminated bamboo lumber (LBL) is considered as a promising green engineering material in civil engineering, due to higher strength/weight ratio and more rapid renewability than wood. An novel OSB webbed bamboo I shaped beam consisted of LBL flanges and OSB web, which were connected by adhesive and nails. The effect of web hole on the performance of the beams were tested to failure to investigate the failure modes and failure mechanism of them. It was showed that the presence of openings has adverse effect on the load carrying capacity and stiffness of composite beams. Holes changed the force transfer path in the web and reduced the strength and stiffness noticeably due to the high shear stresses appearing near the holes. With increasing of hole size, the load carrying capacity of beams decreased dramatically, however the influence was relatively little on the bending stiffness.

1 INTRODUCTION

As environmental protection and emission standards become increasingly strict, environmental-friendly engineering materials came into being, in which bamboo is getting more and more attention. Bamboo is regarded as one of fastest-growing plants on the earth, widely distributed in Africa, South America and Southeast Asia. Especially in southern China, the planting area and output of bamboo ranks first in the world (Xiao 2014, Xiao 2017, Li 2016). Compared with wood, bamboo is stronger, and its strength-to-weight ratio is greater (Xiao 2010, Xu 2014). For decades, raw bamboo material has been successfully used in civil construction, often in the form of beams, columns, rafters and so on (Chung 2002, He 2015). However, several problems were found in the process of practical applications and need to be solved, such as thin-walled hollow, irregular shape, easy to crack when exposure to moisture alternation frequently. Moso bamboo (Phyllostachys pubescens), aged at 4–5 years old, was used to make laminated bamboo lumber (LBL) in Guangdong Province. Analogous to glulam, the original defects of bamboo were dispersed or removed during the production process of LBL (Mahdavi 2012, Huang 2013), so it is thought to be a promising material to replace slowing hardwood in the future.

In recent decades, wood I joists have been commonplace in industrial buildings and civil buildings, especially in low-rise residential buildings. However, the wood I joists are prone to collapse suddenly, which is not conducive to protect dwells from injury of damage (Afzal 2006, Morrissey 2009, Bouldin 2014).

Prefabricated Oriented strand board (OSB) webbed LBL beam, which can be used as a possible alternative to wood I-joist and sawn lumber beams, was introduced. Openings are usually needed to allow services like plumbing, sewage pipes and ventilation systems to run through the web of beam. It is obvious that the holes change the load transfer path in the webs, and then bring disadvantageous effects to shear strength, stability and stiffness.

Considering this, eighteen beams were tested to failure to study the failure modes and failure mechanism of them.

2 MATERIALS AND METHODS

The properties of OSB and LBL were obtained from tension and compression tests. The strengths of OSB in tension/compression were 12.1/13.6, 9.7/12.5 and 10.9/11.3 MPa along the longitudinal, diagonal (45°) and the transverse directions, respectively (Chen, 2017). The MOE and Poisson's ratio of OSB loaded in tension was equaled to that of in compression, approximately 3,000 MPa and 0.2 respectively. The tensile strength of LBL was 107.7 MPa, which was twice as much as that of in compression. The MOE/Poisson's ratio of LBL were 11,000 MPa and 0.26 respectively.

An OSB webbed bamboo composite beam with an I-shaped cross section (Figure 1), which consisted of four LBL flanges and one OSB web with outdoor urea-formaldehyde resin adhesive and nails with spacing of 150 mm, was introduced. The top and bottom flanges can take full advantage of the LBL compressive strength and tensile strength. To avoid uneven adhesive coating, it is extremely important to guarantee that the bonding interface between each other was clean and smooth. Dosage of adhesive between the flanges and web was approximately 250 g/m^2, according to the manufacture's suggestions. Then the OSB web were connected to the LBL with 50 mm common nails at spacing of 150 mm along the length of composite beam. According to the former experience of engineering and test data, the completed beams were stored to cure for 10-15 days inside the laboratory at 22°C and a relative humidity of 60% before testing. Three pairs of LBL stiffeners are placed on both sides of the web at the two supports and points of concentrated loads to help eliminate bearing failures. Several parameters were discussed, including beam height, relative hole size with respect to web height, hole shape and so on (Table 1). All composite beams tested was 2440 mm long, with a support-to-support span of 2000 mm.

As shown in Figures 2 to 4, the specimens with/without opening were tested as simple supported beams in three-point bending according to ASTM D5055, aimed at assessing the mechanical performance of specimens. Load was applied by using an actuator of 100 kN capacity. The load was transferred to the specimens by using a rectangular steel plate, and two roller supports carried the reactions. Three laser displacement sensors (LDS) with an accuracy of ±0.1 mm were used to monitor real time displacement at mid-span and two supports. The specimens were loaded continuously to failure at a displacement rate of 2 mm/s. The resulting force-displacement data, the strain, the load at ultimate limit state and a deflection of L/250 (L is the span between supports) were then recorded by a data acquisition system in 0.2 s increments.

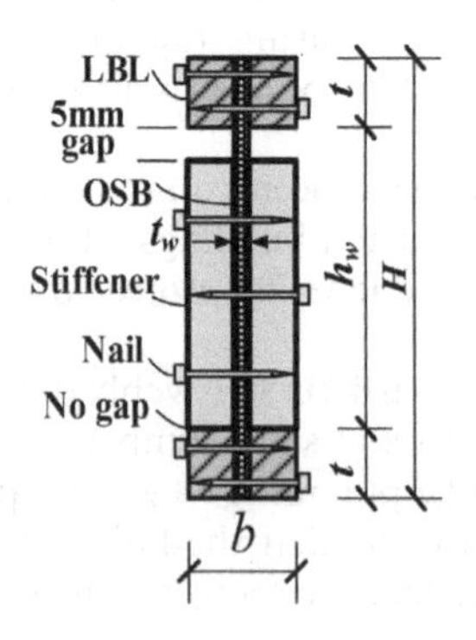

Figure 1. Cross section of specimens.

Table 1. Details of composite beams.

Groups	Beam	hole Shape	b (mm)	t (mm)	H (mm)	d (mm)	h_w (mm)	d/h_w
I	2I1	no hole	59.5	35	240	0	170	0
	3I1	no hole	69.5	35	300	0	230	0
II	2CI1	circle	59.5	35	240	42.5	170	25%
	2CI2	circle	59.5	35	240	85	170	50%
	2CI3	circle	59.5	35	240	127.5	170	75%
	2CI4	circle	59.5	35	240	170	170	100%
	3CI1	circle	69.5	35	300	57.5	230	25%
	3CI2	circle	69.5	35	300	115	230	50%
	3CI3	circle	69.5	35	300	172.5	230	75%
	3CI4	circle	69.5	35	300	230	230	100%
III	2SI1	square	59.5	35	240	42.5	170	25%
	2SI2	square	59.5	35	240	85	170	50%
	2SI3	square	59.5	35	240	127.5	170	75%
	2SI4	square	59.5	35	240	170	170	100%
	3SI1	square	69.5	35	300	57.5	230	25%
	3SI2	square	69.5	35	300	115	230	50%
	3SI3	square	69.5	35	300	172.5	230	75%

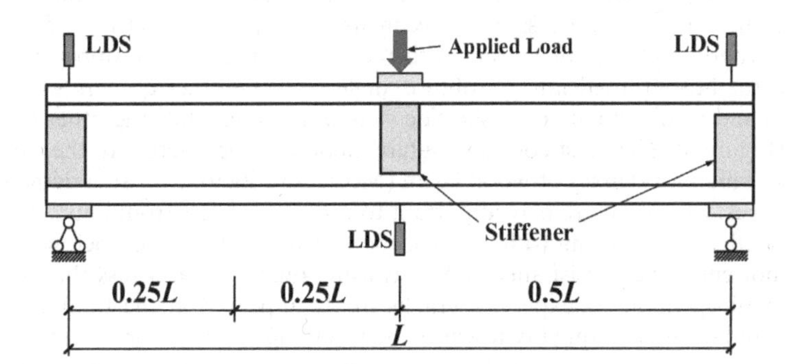

Figure 2. Specimens without opening.

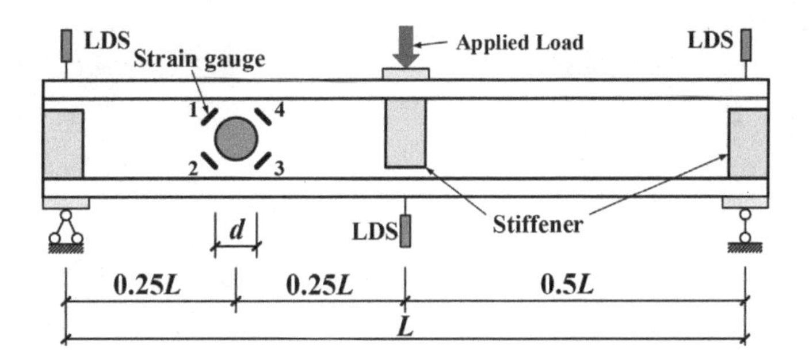

Figure 3. Specimens with circular opening.

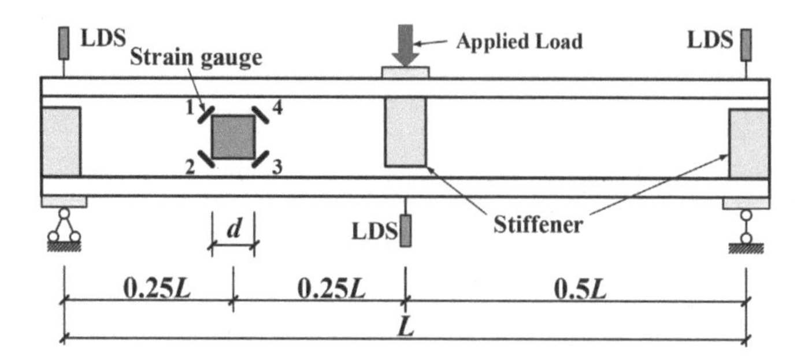

Figure 4. Specimens with square opening.

3 RESULTS AND DISCUSSION

3.1 *Failure mode*

Three failure modes were different for varying hole shape and hole size. Earlier in the testing, for the control specimens (2I1 and 3I1) and specimens with small hole (d/h ≤ 25%) exhibited similar whole damage process of crack initiation and propagation. As the load increased gradually, damage of specimens without opening started from the lateral torsion, accompanied by loud sound. Until the load was reached to about 70 percent of the peak load, the wood flakes of OSB in top flange and bottom flange delaminated slowly along with a sharp crack and propagated in the beams length direction (Figures 4, 5). However, no visible damage in flanges was observed, which was conducive to protect dwellers from injury or damage.

For specimens with hole in the web, the failure mode of composite beams was influenced by the present of hole. The increasingly adverse effects of openings on strength and deformation behavior of beams became more and more obvious with the increasing of hole size. The transfer paths of shear stress in the web was cut off by the opening. A similar failure phenomenon could be observed in the whole test process. For specimens with square holes (d/h > 50%), the square holes became progressively parallelogram with the load increased. The upper right and lower left corners of the holes turned into an obtuse angle, and the other corners were acute angle. The top region and its diagonally opposite side were in tension, while the other regions were in compression (Figure 6). The most common failure mode was the fracture of the web. Unlike the sudden rupture failure of flanges of wood beam (Morrissey, 2009), no visual damage was found in the LBL flanges, which is extremely important to protect dwellers from injury.

For beams with large opening (d/h ≥ 75%), the rest of the OSB web below and above the opening was not enough to resist shear force, causing higher shear stress than the composite beams with small opening. As expected, cracks firstly appeared at the boundary of opening with big noise and developed quickly towards the top flange and bottom flange. After that the sudden increase in shear stress along the adjacent parts of the web caused secondary failures

Figure 5. Delamination of OSB in flanges.

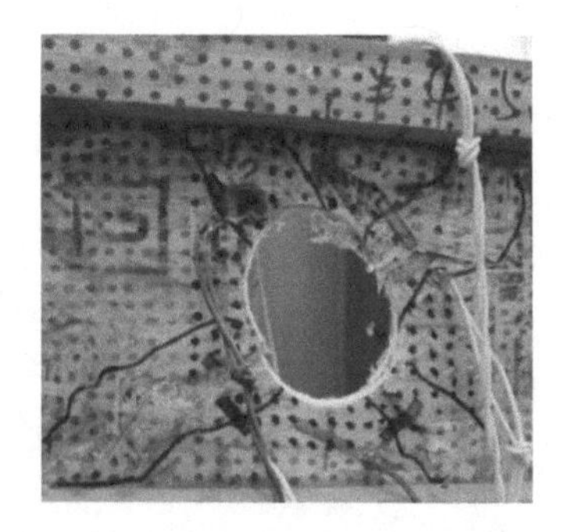

Figure 6. Web shear failure.

Figure 7. Web pullout from flanges.

of web pullout (Figure 7). Surprisingly, no OSB webbed bamboo I shaped beams collapsed, which was very common for wood beams (Morrissey, 2009).

3.2 *Load-deflection behavior*

As shown in Figure 8 and Figure 9, the load-displacement curves were plotted for OSB webbed bamboo beams with and without holes respectively. The strength and bending stiffness of beams with smaller holes (d/h ≤ 25%) were slightly lower than the control beams

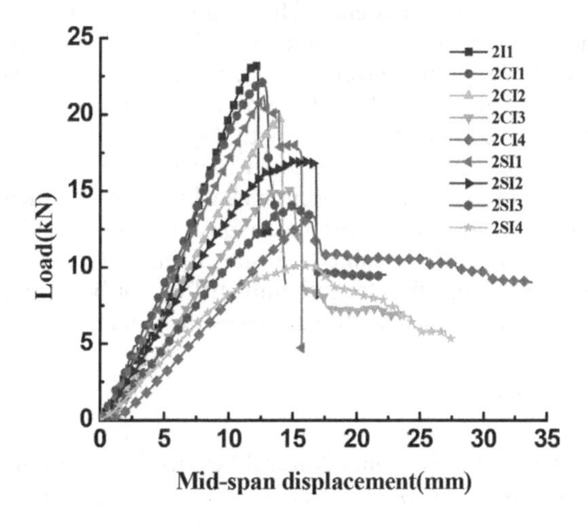

Figure 8. Load - displacement curves for 240 mm depth joists.

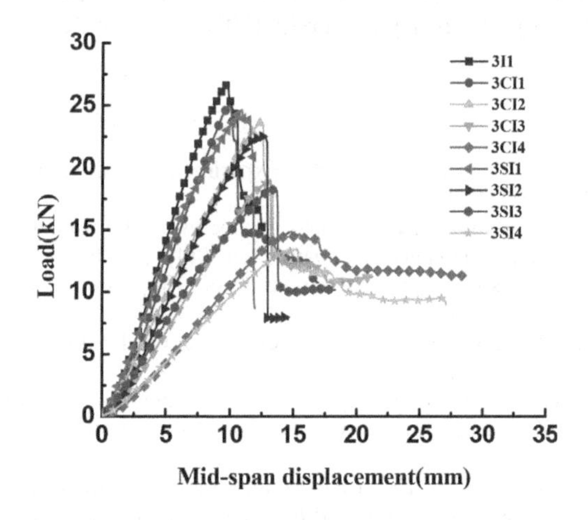

Figure 9. Load - displacement curves for 300 mm depth joists.

without opening, the presence of hole in the web can be ignored. For composite beams with oversize opening, the effect of hole could lead to significant reduction in mechanical performance, especially for beams with big opening (d/h ≥ 75%). Commonly, the failure procedure of specimens can be divided into three phases. The beams worked quite well before reached to the serviceability limit state, followed by a sudden drop, which was associated with hole shape and hole size. Once the peak load of beams was reached, the load carrying capacity of the beams fell to 50% of the peak load suddenly. This explained that the specimens could continue to resist load after the specimens reached ultimate capacity. It must be noted, however, that the residual bearing capacity of specimens differed for varying opening. And the residual strength of beams was closely related to hole size. It was appropriate that the strength and stiffness of beams with a square web hole can be treated the same as a circular hole that inscribes the square hole.

4 CONCLUSIONS

The mechanical properties of OSB webbed bamboo I shaped beams with/without hole in the web were tested and discussed. It was shown that the composite beams exhibited characteristic of brittle failure. Once the peak load of beams was reached, the load carrying capacity of the beams fell to 50% of the peak load suddenly. The strength and stiffness of beams were negatively impacted by the opening in the web, and the adverse effects were more severe as the hole diameter increased. For small hole (d/h ≤ 25%), the effect of the hole on mechanical performance of beams could be ignored. Shear failure of the web around the holes was the primary failure mode for the composite beams with big hole (d/h ≥ 50%). Cracks initially appeared around the web opening and propagated towards the top and bottom flanges. No visible damage was found in the flanges of specimens, which was conducive to protect dwellers from injury or damage. A square web hole has a bigger negative effect on the mechanical behavior than a circular hole with the same hole size, resulted from more serious stress concentration. It was appropriate that the strength and stiffness of beams with a square web hole can be treated the same as a circular hole that inscribes the square hole.

ACKNOWLEDGMENTS

The project was supported by the National Natural Science Foundation of China (51408312), the Natural Science Foundation of Jiangsu Province (BK20130982), the Postgraduate Research & Practice Innovation Program of Jiangsu Province (SJCX18_0313), Ministry of Housing and Urban-Rural Development of China (2018-K5-003), and College Students Practice and Innovation of Nanjing forestry university (201810298055Z).

REFERENCES

Afzal, M.T., Lai, S., Chui, Y.H., & Pirzada, G. 2006. Experimental evaluation of wood I-joists with web holes. *Forest Products Journal* 56(10): 26–30.

Bouldin, J.C., Loferski, J.R., & Hindman, D.P. 2014. Inspection of I-Joists in Residential Construction. *Practice Periodical on Structural Design and Construction* 19(4): 04014016.

Chen, G., & He, B. 2017. Stress-strain constitutive relation of OSB under axial loading: *An experimental investigation. Bioresources* 12(3): 6142–6156.

Chung, K.F., & Yu, W.K. 2002. Mechanical properties of structural bamboo for bamboo scaffoldings. *Engineering Structures* 24(4): 429–442.

He, M., Li, Z., Sun, Y., & Ma, R. 2015. Experimental investigations on mechanical properties and column buckling behavior of structural bamboo. *Structural Design of Tall and Special Buildings* 24(7): 491–503.

Huang, D., Zhou, A., & Bian, Y. 2013. Experimental and analytical study on the nonlinear bending of parallel strand bamboo beams. *Construction and Building Materials* 44(8): 585–592.

Li, H.T., Wu, G., Zhang, Q.S., & Su, J.W. 2016. Mechanical evaluation for laminated bamboo lumber along two eccentric compression directions. *Journal of Wood Science* 62(6): 1–15.

Mahdavi, M., Clouston, P.L., & Arwade, S.R. 2012. A low-technology approach toward fabrication of Laminated Bamboo Lumber. *Construction and Building Materials* 29(4): 257–262.

Morrissey, G.C., Dinehart, D.W., & Dunn, W.G. 2009. Wood I-Joists with Excessive Web Openings: An Experimental and Analytical Investigation. *Journal of Structural Engineering* 135(6): 655–665.

Xiao, Y., Zhou, Q., & Shan, B. 2010. Design and Construction of Modern Bamboo Bridges. *Journal of Bridge Engineering* 15(5): 533–541.

Xiao, Y., Chen, G., & Feng, L. 2014. Experimental studies on roof trusses made of glubam. *Materials and Structures* 47(11): 1879–1890.

Xiao, Y., Wu, Y., Li, J., & Yang, R.Z. 2017. An experimental study on shear strength of glubam. *Construction and Building Materials*. 150: 490–500.

Xu, Q., Harries, K., Li, X., Liu, Q., & Gottron, J. 2014. Mechanical properties of structural bamboo following immersion in water. *Engineering Structures* 81: 230–239.

Modern Engineered Bamboo Structures – Xiao, Li & Liu (eds)
© 2020 Taylor & Francis Group, London, ISBN 978-1-138-35185-1

Experimental study on bending behavior of CLB panels

H. Ding & Q.F. Lv
Southeast University, Nanjing, Jiangsu, China

ABSTRACT: Cross-laminated bamboo (CLB) is an innovative engineering bamboo panel product made from gluing layers of solid-sawn bamboo at perpendicular angles. Owing to the excellent structural rigidity in both orthogonal directions, CLB becomes a preferred construction material for shear walls, floor diaphragms and roof assemblies. The bending property of CLB panels were investigated and the effects of thickness of panels and numbers of layers on bending behavior were compared in this paper. It shows that CLB panels have strong flexural bearing capacity. Under the same conditions, the larger the thickness of panels, the more the number of layers, the better the bending behavior.

1 INTRODUCTION

Bamboo plants play a very important role in the world's forest resources and are honored as the "second largest forest in the world" (Yingshuang et al., 2008). The quantity and quality of bamboo resources in China are in the forefront of the world. Bamboo is a sustainable and environmental material in architecture industry with good mechanical properties. The tensile strength of bamboo is about twice and the compressive strength is 1.5 times of the wood. At the same time, demand for wood structural profiles in our country is increasing, while the timber resources are relatively scarce. It is important to accelerate the development of bamboo structural profiles.

Yan Xiao in Hunan University tested the mechanical behavior of Glubam and obtained the compressive strength, elastic modulus, bending strength and shear strength of Glubam (Yan et al., 2012).

A kind of cross-laminated gluing bamboo material is presented in this paper. It is called Cross-laminated bamboo (CLB). CLB is a massive engineering bamboo product made by gluing cross-wise layers of sawn bamboo to form large-scale timber panels. The cross section of a CLB panel usually consists of three to seven glued layers of dimension bamboos placed in orthogonally alternating orientation to the neighboring layers. CLB has advantages of similar mechanical properties in orthogonal directions, large in-plane strength and stiffness, and high level of prefabrication, etc.

CLB has the same structure as CLT(Cross-Laminated Timber) material and some similarity in performance. Therefore, it has certain reference significance.

Han-Min Park et al. tested the static bending behavior of CLT and compared with the traditional glulam structure (Park et al., 2003). The results showed that the elastic modulus and bending strength of CLT were better than glulam under the same conditions. Johannes and Schneide studied the CLT as a solution for medium and high-rise building materials. The dimensional stability, seismic resistance and fire resistance of CLT shear wall were tested and analyzed (Schneider et al., 2012). Zhiwei Zhang and Meizhen Fu analyzed the influence of CLT and reinforced concrete on environment. It turns out that the impact of CLT on environment is only 51% of the same volume of reinforced concrete. The advantage of CLT is more obvious in terms of reducing the impact of greenhouse (Zhiwei et al., 2014). Zeya Liu in Central South University of Forestry and Technology studied the mechanical behavior of larch glued wood through three-point bending and four-point bending experiments. The bending property of CLB panels was investigated. The results showed that the stress distribution of specimens basically conformed to the assumption of flat section (Zeya et al., 2015).

2 MATERIALS AND METHODS

2.1 *Materials and specimens*

The specimens are divided into three groups: CLB-A、CLB-B and CLB-C. The length of the layer used in the longitudinal lamination was 1800 mm and that of the layer used in the perpendicular laminations was 600 mm . The details are in Table 1.

CLB-A: The nominal cross-section dimension of the layer was 100 mm × 20 mm. The CLB-A panels were assembled in cold press, each panel was made up of five orthogonally crossed layers, and the thickness of each CLB-A panel was 100 mm.

CLB-B: The nominal cross-section dimension of the layer was 100 mm × 20 mm and 100 mm × 12 mm. The CLB-B panels were assembled in cold press, each panel was made up of seven orthogonally crossed layers. The thickness of serface layer was 20 mm, and the thickness of interlayer was 12 mm. The thickness of each CLB-B panes was 100 mm.

CLB-C: The nominal cross-section dimension of the layer was 100 mm × 12 mm. The CLB-C panels were assembled in cold press, each panel was made up of five orthogonally crossed layers, and the thickness of each CLB-C panel was 60 mm.

2.2 *Material properties*

Referring to *General principles of physical and mechanical test methods for wood* (GB/T 1928-2009), mechanical properties of bamboo layers were tested. Test results are shown in Table 2.

Table 1. Details of specimens.

Type of specimens	Size of specimens (length × width × thickness)/mm	Details of specimens	Number of specimens
CLB-A	1800 × 600 × 100	5 layers, thickness of each layer is 20 mm	5
CLB-B	1800 × 600 × 100	7 layers, thickness of serface layer is 20 mm and the interlayer is 12 mm	5
CLB-C	1800 × 600 × 60	5 layers, thickness of each layer is 12 mm	5

Table 2. Mechanical properties of bamboo layers.

Mechanical properties	Type of layer	12 mm	20 mm
Parallel-to-grain direction, tension	Strength/MPa	47.45	
	MOE/MPa	17949	
Perpendicular-to-grain direction, tension	Strength/MPa	7.86	
	MOE/MPa	3751	
Parallel-to-grain direction, compression	Strength/MPa	108.87	107.99
	MOE/MPa	20769	21548
Perpendicular-to-grain direction, compression	Strength/MPa	45.90	51.64
	MOE/MPa	5979	6368
Parallel-to-grain direction, bending	Strength/MPa	94.21	88.10
	MOE/MPa	10657	10086

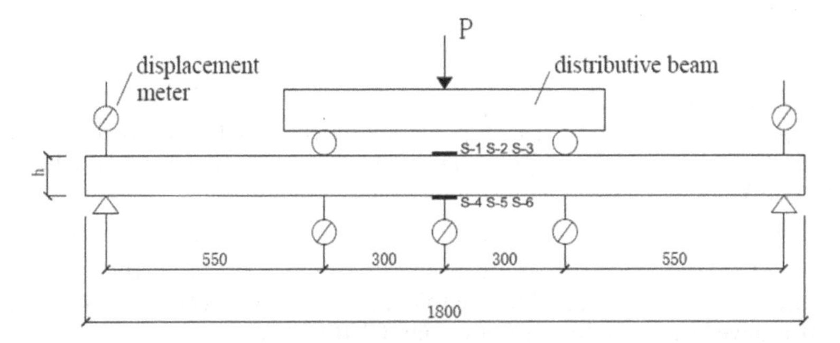

Figure 1. Loading configurations of the CLB bending specimens.

2.3 *Test program*

The method of 4-point loading was used in the test. Before the formal loading, a 10 kN preload was applied on the CLB specimens, eliminating the impact of inelastic deformation and checking if the equipment is working properly. After preloading, a constant loading which was performed by force control at 5 kN per level was applied on the CLT specimens. Loading and position of strain gauge and displacement gauge were shown in Figure 1.

The data to be measured during the test are below:

Strain and deflection of the span-center of bamboo layer in CLB panels was recorded. The resistance strain gauge were arranged on the top, bottom and side of the CLB panel. The dialgauges were placed under the supports, loading points and span-center. The deflection of the span-center was calculated by vertical displacement difference of each gauge.

3 RESULTS AND ANALYSIS

3.1 *Experimental phenomenon*

At the initial stage of loading, there is no other phenomenon except deflection gradually increasing. When the load exceeds the estimated proportional limit, the deflection continues to increase. The CLB panel have obvious bending while the loading is continuing, which indicate that the CLB panel has good deformation capacity. When approaching the ultimate failure load, there are some cracks under the loading point. As the load increases, the crack width was increasing gradually. When the load increases to the ultimate failure load, the span-center fiber of bottom layer was pulled off in the first place. The failure mode pictures of specimens are shown in Figure 2.

Figure 2. CLB panels failure mode.

3.2 *Results and analysis*

3.2.1 *Load-displacement curve*

The middle span deflection of 5 CLB-A panel specimens, 5 CLB-B panel specimens and 5 CLB-C panel specimens were analyzed. The load-displacement curve of CLB panels are shown in Figure 3.

From Figure 3, we can see:

(a) The curves of CLB-A and CLB-B are steeper than the curve of CLB-C. The thickness of CLB-A and CLB-B is 100 mm while the thickness of CLB-C is 60 mm, as a result, CLB-A and CLB-B have larger bending stiffness than CLB-C.

(b) The curve of CLB-B is a little steeper than the curve of CLB-A. CLB-B panels are exactly the size and thickness of CLB-A panels. But CLB-B panels have more layers. Therefore, CLB-B panels have more ratio of longitudinal lamination and have larger bending stiffness than CLB-A.

3.2.2 *Load-strain curve*

Top and bottom strain of the cross section of specimens under different loads are shown in Figure 4-6. S-1, S-2, S-3 are strain gauges attached on the top of the specimens. S-4, S-5, S-6 are strain gauges attached on the bottom of the specimens.

The strain of each section of specimen increases gradually with the increase of load and the pattern of change is close. As the load increases, the rate of strain change increases slightly.

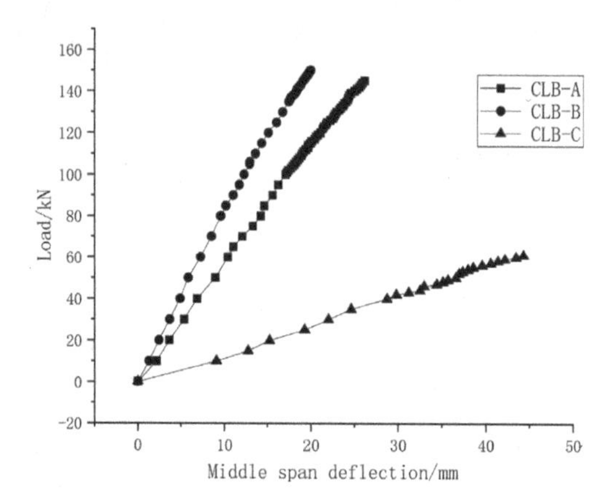

Figure 3. Load-deflection curve.

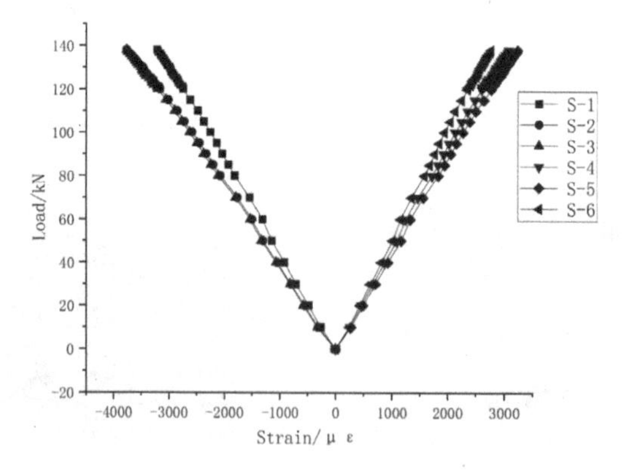

Figure 4. Load-strain curve of CLB-A.

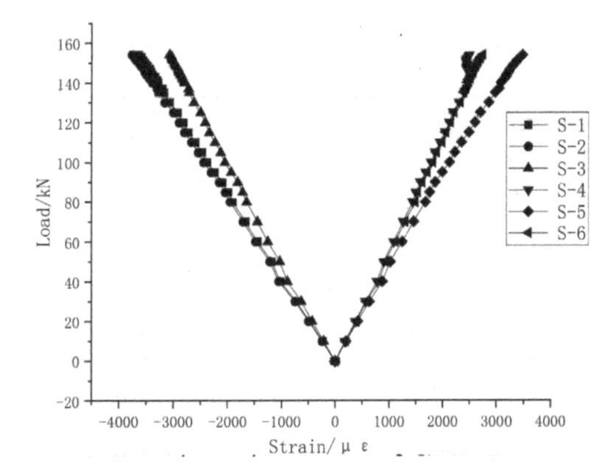

Figure 5. Load-strain curve of CLB-B.

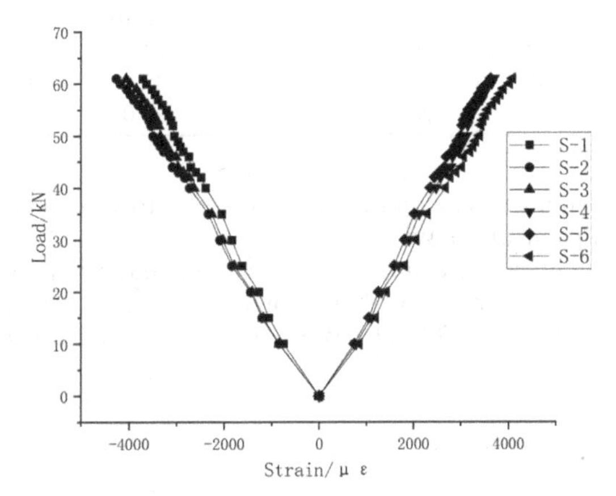

Figure 6. Load-strain curve of CLB-C.

Hypothetically, the specimen remains completely linear in stress and strain. The pure bending stress of the panel is $\sigma = E\varepsilon = \frac{My}{I}$, than we can obtain: $\varepsilon = \frac{My}{EI}$. So the growth rate of strain is inversely proportional to effective bending stiffness EI. That is to say, when the load is small, the stiffness of the specimen is good. As the load increases, the overall stiffness of the specimen decreases slightly. The effect of interlaminar combination is weakened, but the attenuation is small. It indicates that the integrity of the specimen is good, and the change of overall stiffness is very small.

3.2.3 *Mechanical analysis*

Hypothetically, the specimen remains completely linear in stress and strain. The pure bending stress of the panel is $\sigma = E\varepsilon = \frac{My}{I}$, than we can obtain: $\varepsilon = \frac{My}{EI}$. So the growth rate of strain is inversely proportional to effective bending stiffness EI. That is to say, when the load is small, the stiffness of the specimen is good. As the load increases, the overall stiffness of the specimen decreases slightly. The effect of interlaminar combination is weakened, but the attenuation is small. It indicates that the integrity of the specimen is good, and the change of overall stiffness is very small.

Hypothetically, the specimen remains completely linear in stress and strain. The middle span stress at bottom of panel is $\sigma = \frac{M}{W_Z} = \frac{\frac{1}{6}PL}{\frac{1}{6}bh^2} = \frac{PL}{bh^2}$. The middle span stress at bottom of panel from the experiment is $\sigma = E\varepsilon$. The theoretical value of the middle span stress at the

Table 3. The middle span stress at the bottom of each plate at failure in theory and from experiment.

Test panel	Failure load/kN	Middle span strain/$\mu\varepsilon$	σ_T	σ_E	Reduction coefficient
A1	137	2880	38.82	29.05	25.17%
A2	145	3386	41.08	34.15	16.87%
A3	138	3180	39.10	32.07	17.97%
A4	138	3235	39.10	32.63	16.55%
A5	136	3598	38.53	36.29	5.82%
B1	150	2110	42.50	21.28	49.93%
B2	154	3486	43.63	35.16	19.42%
B3	143	2358	40.52	23.78	41.30%
B4	126	1407	35.70	14.19	60.25%
B5	148	3363	41.93	33.92	19.11%
C1	62	1853	48.80	19.75	59.53%
C2	52	1801	40.93	19.19	53.10%
C3	59	2118	46.44	22.57	51.39%
C4	61	3623	48.01	38.61	19.58%
C5	56	3290	44.07	35.06	20.45%

bottom of each plate at failure (σ_T), and the middle span stress at bottom of panel from the experiment (σ_E) of each panel are shown in Table 3.

Due to the uncertainty of the experiment, the maximum and minimum value of each group was removed when the average value of Reduction coefficient was calculated. The calculated results are below: CLB-A panel: 17.13%, CLB-B panel: 36.88%, CLB-C panel: 41.65%.

4 CONCLUSION

In this paper, an experimental study is conducted to investigate the flecural behavior of three kinds of CLB panels, and the load-displacement curve and load-strain curve of test panels are obtained.

It can be seen from the test results, the failure modes of three kinds of CLB panels are basically the same. All were broken by the span-center fiber of bottom layer pulled off. CLB panels have good bending resistance, which completely satisfy the requirement of roofs and floors in Chinese architectural structure load standards.

The CLB-A and CLB-B have the same thickness, but CLB-B has more layers and more ratio of longitudinal lamination than CLB-A. Therefore, CLB-B has better flexural rigidity and larger bearing capacity than that of CLB-A. The CLB-A and CLB-c have the exact number of layers, but the thickness of layer and the whole panel of CLB-A is larger than that of CLB-C. Therefore, CLB-A has better flexural rigidity and larger bearing capacity than that of CLB-C.

ACKNOWLEDGEMENT

The authors gratefully acknowledge the support from National Key Research and Development Plan of China (Grant NO. 2017YFC0703500).

REFERENCES

Park H.M., Fushitani M., Sato K., et al. 2003. Static bending strength performances of cross-laminated woods made with five species. *Journal of Wood Science* 49(5): 411–417.

Schneider J, Stiemer S.F, Tesfamariam S, et al. 2012. Damage assessment of cross laminated timber connections subjected to simulated earthquake loads.

Y. Xiao & R.Z. Yang 2012. Experimental research on mechanical properties of glubam. *Jorunal of Building Structures* 33(11): 150–157.
Yingshuang B. & Zhiyong L. 2005. Development status and trend of foreign bamboo industry. World Bamboo and Rattan Communication 3(4): 40–42.
Zeya L. 2015. Experimental research on mechanical properties of larch glued wood. Central South University of Forestry and Technology.
Zhiwei Z. et al. 2014. Comparative analysis of the environmental impact of CLT and reinforced concrete materials. National Bridge Academic Conference.

Modern Engineered Bamboo Structures – Xiao, Li & Liu (eds)
 ISBN 978-1-138-35185-1

Flexural behavior of glubam-recycled aggregate concrete composite beam

S.T. Deresa
Nanjing Tech University, Nanjing, Jiangsu, China
Addis Ababa Science and Technology University, Addis Ababa, Ethiopia

H.T. Ren & J.J. Xu
Nanjing Tech University, Nanjing, Jiangsu, China

ABSTRACT: This paper presents experimental results of bending test of composite beam made of recycled concrete slab and glue-laminated bamboo (glubam) beam. Two types of shear connectors, which are commonly used in timber concrete composite (TCC) beams, were used for the test. The load/mid-span displacement curve, strain distribution curve at the mid-span section of the composite beam, and the load-relative slip curve of the composite beam were obtained. The relevant mechanical properties such as the maximum bearing capacity and the flexural stiffness of the composite beam are also obtained. Analysis and comparison of the results show that the failure mode of the composite beam with screw-connector is a delamination cracking failure of the glued bamboo beam while the beam with notch connection showed a flexure tensile failure. And use of recycled aggregate concrete is viable for bamboo-concrete composite beam.

1 INTRODUCTION

Sustainability is becoming major concern in construction industry owing to, on one hand, scarcity of natural resources, on the other hand, environmental impacts resulted from waste generated and the way how materials are sourced. Utilization of renewable natural resources as a source of material is an alternative leading to a sustainable construction. In this regard, the benefit of employing recycled aggregate and bamboo in construction is enormous in terms of minimizing both environmental impact and depilation of scarce natural resources.

Recycling of waste is a promising technology for the use of recycled aggregate (RA) in concrete. Reinforced recycled aggregate concrete (RRAC) have been received a comparable performance against normal RC structures made of natural aggregates (Marthong, Sangma, Choudhury, Pyrbot & Bharti, 2017; Senaratne, Lambrousis, Mirza, Tam & Kang, 2017).

Bamboo, a rapidly renewable and sustainable resource, has many advantages as a construction material. Currently, there is a growing attention towards development of bamboo products as a sustainable, cost effective and eco-friendly construction material. Glubam, engineered bamboo made by laminating bamboo fibers with adhesives, can be used as a structural element and has mechanical properties that are comparable or higher than timber (Xiao, Shan, Chen, Zhou & She, 2007; Xiao, Yang & Shan, 2013). It also has acceptable durability in exposure to outdoor conditions with modest remedial measures (Shan, Chen & Xiao, 2012).

Proper understanding of bending properties of bamboo is important for its wider application in structural use. When compared to reinforced concrete and steel structures, glubam-only flexural members exhibit a relatively low stiffness due to the inherent flexibility of bamboo (Shan, Xiao, Zhang & Liu, 2017; Sharma et al., 2015). To improve the bending performance of glubam members is therefore an important step forward to better utilization of the glubam members for structural use. (Shan et al., 2017) developed a new type of composite structure called bamboo-concrete composite (BCC) system aiming to improve the bending performance of Glulam-only members. They adopted the same connection technologies that are used in timber-concrete composite (TCC) beams (Auclair, Sorelli & Salenikovich, 2016; Ceccotti, Fragiacomo & Giordano, 2007; Eisenhut, Seim & Kühlborn, 2016; Gutkowski, Brown, Shigidi & Natterer, 2008). TCC beams are timber-concrete composite which are made of lower timber beam and upper concrete slab and connected with reliable shear connectors and designed in such a way that the timber carries the tensile force and the concrete resists the compression to take an optimized benefit of the properties of the two materials.

This paper reports experimental results of static bending test of glubam-recycled concrete composite beam (BRACC) using screw and notched shear connectors, which have been used in the above-mentioned composite systems (TCC & BCC).

2 EXPERIMENTAL PROGRAM

2.1 *Material*

The glubam beam was made of two pieces of each 30 mm thick laminate. The connectors were glued in glubam beams by a two-component epoxy adhesive provided by a local manufacture. Mechanical properties of glubam and adhesive are presented in Table 1-2. Ordinary Portland cement, river sand as natural fine aggregate, 70% natural coarse aggregate and 30% recycled coarse aggregate (proportioned by volume) were used for manufacturing the concrete slab. The mix proportion for the concrete mix was 1.0: 0.40: 1.08: 2.40 (cement: water: sand: gravel). Concrete was designed for target compressive strength of 40 MPa.

2.2 *Test specimen*

Two 3600 mm long BRACC beams, and one glubam-only beam were prepared for the test. The size of concrete slab was 3600 mm × 1000 mm × 100 mm (length × width × thickness). The glubam bamboo beam was made of two pieces of each 30 mm thick and 3600 mm long laminate. Two types of connection system, notch (NC) and screw (SC) connections were used. A screw connector (SC) used was 18 mm diameter and 180 mm long with a nut welded on the top. 100 mm deep and 20 mm diameter holes were drilled at 100 mm spacing for shear span and 200 mm spacing in the span between the loading points on the glubam beam, then adhesives filled and the screw inserted in the predrilled holes. In the notched connection (NC), a notch with 100 mm length and 50 mm depth was cut in the glubam beam. The notches are spaced at 250 mm center to center in the shear span and 300 mm in the span between the loading points. A screw, the same type used in the screw connector, was used along with the notch

Table 1. Mechanical properties of glubam specimens (parallel to grain, MPa).

Compressive strength (main bamboo fiber direction)	77
Tensile strength (main bamboo fiber direction)	90
Shear strength	4.5
Elastic modulus	10400

Table 2. Mechanical properties of two-component epoxy resin.

Specific gravity of mix (g/cm^3)	1.5± 0.1
Compressive strength (MPa)	90.9
Tensile bending strength (MPa)	79.5
Splitting strength (MPa)	10.4

connection and glued to the beam in the same way as it was done in the screw connection but only for a depth of 50 mm. The parameters used in specimen fabrication and connection details are given in Table 3 and Figures 1–2, respectively.

2.3 *Flexural test setup and procedure*

The flexural performance was evaluated with a four point load bending test on a simply supported bamboo-recycled aggregate composite beam. Figure 3 shows the experimental test set up and loading device scheme.

The beams were supported by roller and pin support with a clear span of 3500 mm and a flexural span of 1200 mm. The load was applied in three phases. The first phase was loading with load control. At each loading interval, the specimen was loaded with load increment of 5 kN. It continued until the displacement in the mid-span reached L/300, load at which the specimen was considered to have reached serviceability limit state (SLS). After this, the second loading phase starts. The second loading phase were again loading with load control, but the load increment after each loading interval was 2.5 kN. This continued until the displacement in the mid span of the composite beam reached L/150, where the specimen was considered to have reached the Ultimate limit state (ULS). At the final stage, the loading continued with displacement control. Each loading interval had 1mm increase in mid-span displacement and the loading continued until failure load was reached. The failure load was assumed to be 85% the ultimate load. Mid-span displacement of the glulam beam was

Table 3. Specimen parameter details.

Test specimen	Specimen ID	Concrete slab type	Concrete slab size (mm)	Connection
Specimen 1	NC	RAC 30*	3600 × 1000 × 100	Notch
Specimen 2	SC	RAC 30*	3600 × 1000 × 100	Screw
Specimen 3	Glubam-only Beam	-	-	-

*RAC 30: recycled aggregate concrete with recycled aggregate replacement percentage of 30%.

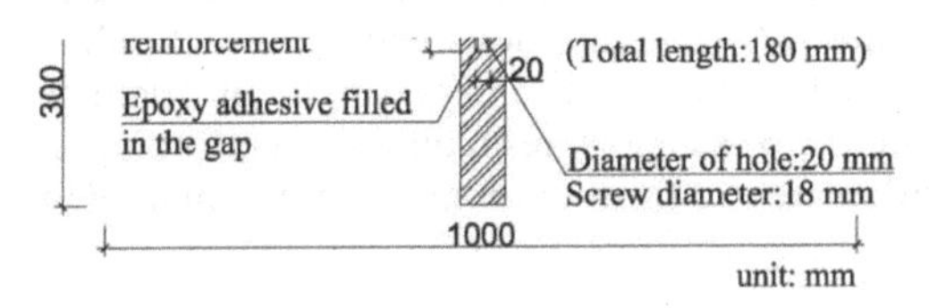

Figure 1. Screw connection detail.

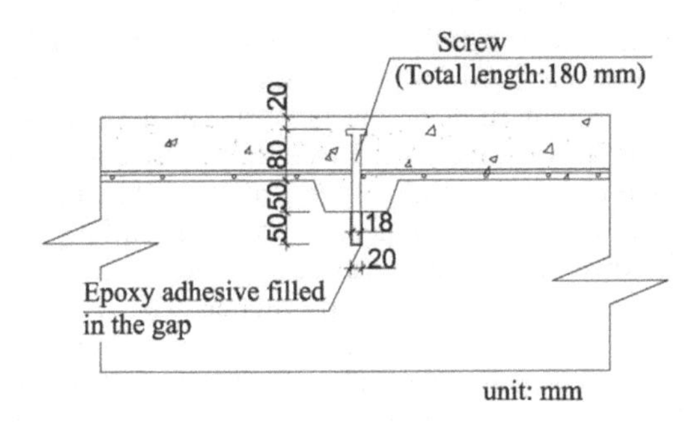

Figure 2. Notch connection details.

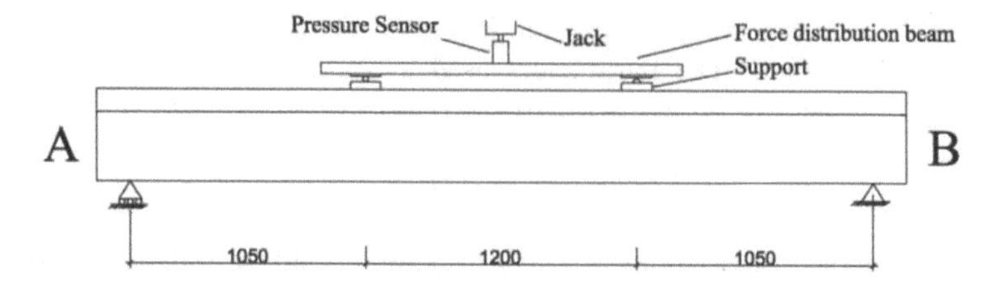

Figure 3. Loading set up and device schematic.

measured using linear voltage displacement transducers LVDTs. Relative slip between the concrete slab and the glulam beam was also measured using LVDTs at different location on the beam.

3 EXPERIMENTAL RESULTS AND DISCUSSION

3.1 *Failure mode*

All specimens exhibited essentially a linear behavior before they reached the load at serviceability limit state (SLS). As the load increases, the composite beam starts to gradually lose its stiffness. Consequently; glubam layers pulled off at lower surface of bamboo beam in the specimen with notch shear connector; while inter-layer cracking of the bamboo beam was observed for the specimen with screw shear connector. As the load come near to the peak, the glued bamboo in the composite beam with notch shear connector (NC) breaks in tension at the bottom in the span between the loading points. And it extends along the length of the beam up on increasing the load and finally the beam fails in tensile failure mode. In the composite beam with screw shear connector (SC), beam failed by delamination cracking, where the glubam beam cracked and splitting of the glubam layers along the longitudinal direction of bamboo beam occurs. No significant damage was observed on the concrete slab, it has cracked dominantly in the span between the loading points. Small cracks were initiated before the load reached the ultimate limit capacity. Up to 4mm crack width was observed at failure for specimen with notch shear connector (NC). Careful examination of the specimens after complete failure revealed that the screw has given good anchor for the connections with sufficient pull out strength in both connections. This proves that the screw embedment length of 50mm for NC connection and 100mm for SC connection as well as epoxy resin adhesive performance are sufficient to provide the desirable bonding of the screw to form the composite system in both connections.

3.2 *Load bearing capacity and Load/mid-span displacement curve*

Experimental results of load bearing capacity, mid-span displacement and efficiency of connections at SLS and ULS are given in Table 4 below. Load/mid-span displacement curves are presented in Figure 4. The experimental curves are presented within two extreme stiffness limit curves, the fully composite and fully non composite curves. These limiting carves are drawn based on the European standard for timber design, Eurocode 5 (DIN 1994). The transformed section method and summation of the flexural stiffness values of the concrete and glubam obtained from material tests were used to obtain fully composite and fully non composite curves, respectively (Gutkowski et al., 2008; Lukaszewska et al., 2010). As can be seen from the figure 4, Irrespective of the connection type, essentially linear relationship was observed in load/mid-span displacement curve until the load reaches serviceability limit (SLS). Specimen with Screw connection (SC) has slightly higher peak load capacity than the notch (NC) connection. Peak load of 186.9 kN and corresponding mid-span displacement of 24.08 mm was recorded for screw connection. The bearing capacity of glubam-only is also presented for comparison. Connection efficiency was calculated by Equation (1) (Gutkowski et al., 2008).

$$\text{Efficiency} = \frac{D_N - D_I}{D_N - D_C} \times 100 \tag{1}$$

Where D_C is the theoretical fully composite displacement, D_N the theoretical fully non-composite displacement and D_I is the measured displacement for incomplete composite action of the specimen.

Connection efficiency is the extent of the composite action observed as a percentage of maximum possible (Lukaszewska et al., 2010). As can be seen from the table, the calculated connection efficiency show that SC connection has exhibited better efficiency at SLS and ULS load levels.

3.3 *Strain distribution curve at mid-span*

Figures 5–6 show the strain distribution of the composite beam cross-section at mid-span. As shown in the figures, the strain distribution for specimen NC is almost linear before the load reached serviceability limit state (SLS). The difference in flexural strain distribution at the interface of the two materials is small and the combined composite action was better. When the serviceability limit state is exceeded, the difference gradually becomes larger. For SC connection, the difference in flexural strain at the interface of the two materials is relatively small, and the overall combined effect was good. Irrespective of connection type, the neutral axes for all the cross-sectional strain distribution curve are in the same position (at the lower part of the concrete slab), which is the same as the theoretically calculated neutral axis position (the "gamma method" in annex 2 of European code 5). This is evidence that the shear connector achieved a good composite action between the glubam and the recycled concrete slab.

Table 4. Load bearing capacity.

Specimen ID	Peak Load P_{max} (kN)	Maximum mid span deflection (mm)	Efficiency (%) SLS	Efficiency (%) ULS
NC	180	24.57	54	32
SC	186.9	24.08	59	36
Glubam-only	72.9	63.22	-	-

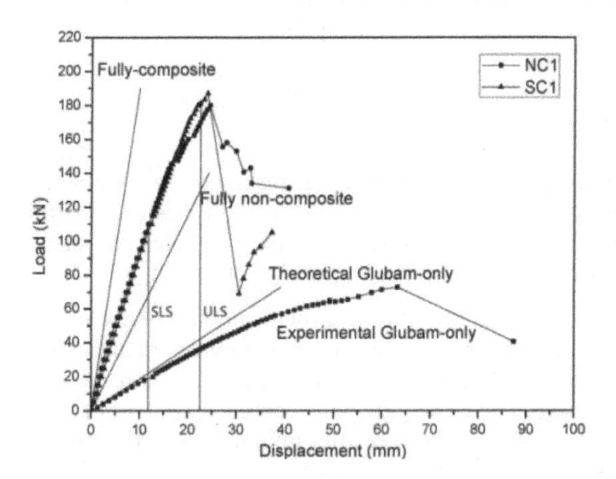

Figure 4. Load/mid-span displacement curves.

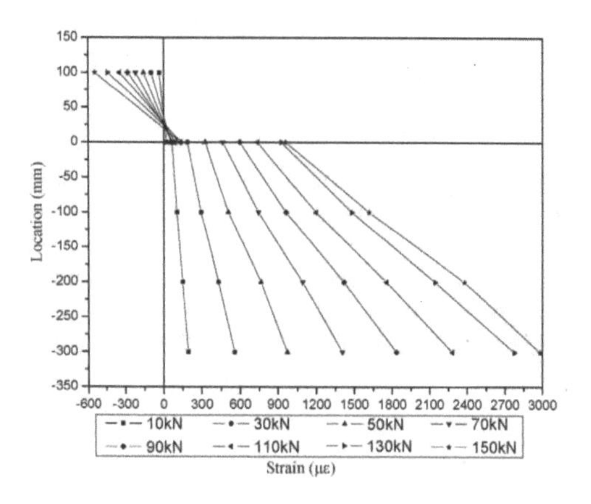

Figure 5. Notch connection (NC).

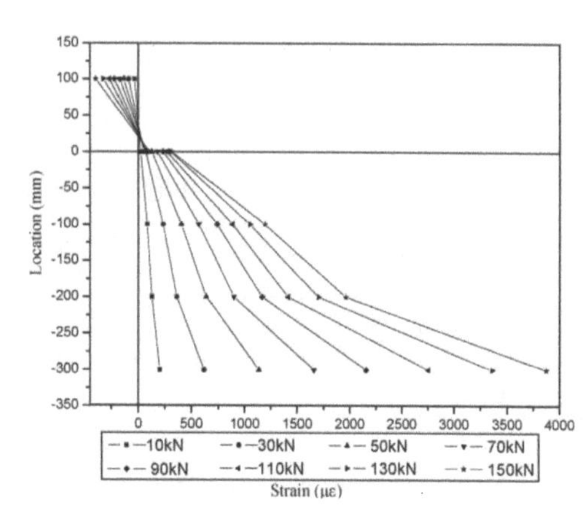

Figure 6. Screw connection (SC).

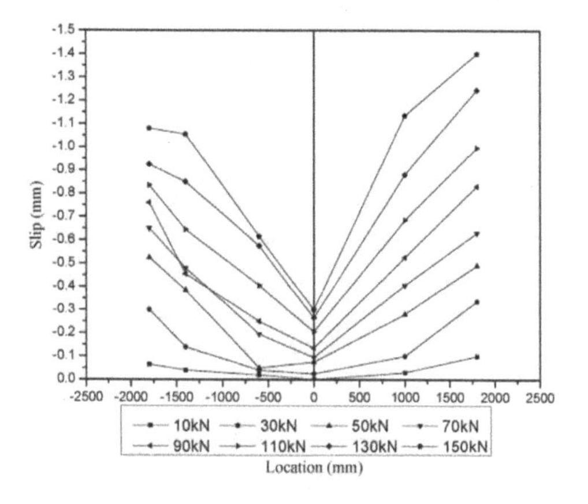

Figure 7. Notch Connection (NC).

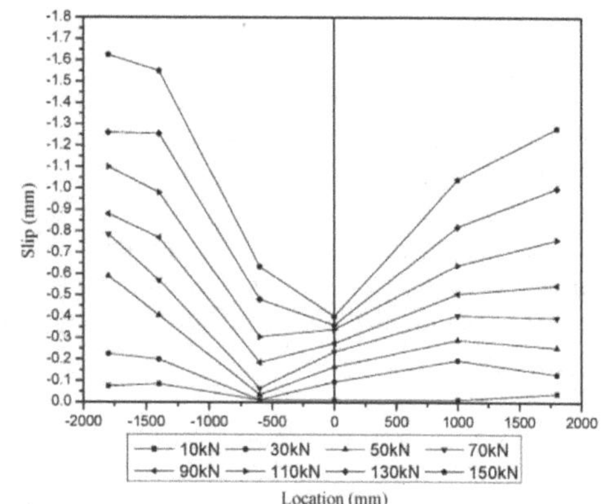

Figure 8. Screw Connection (SC).

3.4 *Load-slip curve*

The relative slip in the interface of the composite beam was measured at the location shown in Figures 7–8. The relative slip is basically symmetrical along the left and right side of the mid span. The slip at the ends are higher than the slip at the interior points, and the relative slip in the mid-span is negligible.

4 CONCLUSIONS

This study reports the experimental bending test results on two glubam-recycled concrete slab composite beam and one glubam-only beam with two types shear connectors to understand the overall flexural behavior of the glubam recycled concrete composite beam. The following conclusions can be drawn from the experimental investigation.

1. Glubam recycled concrete composite beam with notch connection failed in tensile failure mode while the beam with screw connection failed by delamination cracking.
2. The result of the bending test shows that it is viable to use recycled aggregate slab for the composite beam. Use of recycled aggregate concrete slab doesn't affect the failure mode as the concrete had no significant damage after final failure of the composite beam.
3. Load bearing capacity and efficiency of screw connection is slightly higher than the notch connections.

REFERENCES

Auclair, S.C., Sorelli, L. & Salenikovich, A. 2016. A new composite connector for timber-concrete composite structures. *Journal of Construction and Building Materials*, 112: 84–92.

Ceccotti, A., Fragiacomo, M. & Giordano, S. 2007. Long-term and collapse tests on a timber-concrete composite beam with glued-in connection. *Journal of Materials and Structures*, 40(1): 15–25.

Eisenhut, L., Seim, W. & Kühlborn, S. 2016. Adhesive-bonded timber-concrete composites Experimental and numerical investigation of hygrothermal effects. *Journal of Engineering Structures*, 125: 167–178.

Gutkowski, R., Brown, K., Shigidi, A. & Natterer, J. 2008. Laboratory tests of composite wood-concrete beams. *Construction and Building Materials*, 22(6): 1059–1066.

Lukaszewska, E., Fragiacomo, M. & Johnsson, H. 2010. Laboratory Tests and Numerical Analyses of Prefabricated Timber-Concrete Composite Floors. *Journal of Structural Engineering*, 136(1): 46–55.

Marthong, C., Sangma, A.S., Choudhury, S.A., Pyrbot, R.N., Tron, S.L., Mawroh, L. & Bharti, G.S. 2017. Structural Behavior of Recycled Aggregate Concrete Beam-Column Connection in Presence of Micro Concrete at Joint Region. *Journal of Structures*, 11: 243–251.

Senaratne, S., Lambrousis, G. & Kang, W.-H. 2017. Recycled Concrete in Structural Applications for Sustainable Construction Practices in Australia. *Journal of Procedia Engineering*, 180: 751–758.

Shan, B., Chen, J. & Xiao, Y. 2012. Mechanical Properties of Glubam Sheets after Artificial Accelerated Aging. *Journal of Key Engineering Materials*, 517: 43–50.

Shan, B., Xiao, Y., Zhang, W.L. & Liu, B. 2017. Mechanical behavior of connections for glubam-concrete composite beams. *Journal of Construction and Building Materials*, 143: 158–168.

Xiao, Y., Shan, B. & Chen, G. 2007. Development of a new type of glulam-glubam. Modern. Bamboo Structure, *Proceedings of the First International Conference on Modern Bamboo structures*:299.

Xiao, Y., Yang, R.Z. & Shan, B. 2013. Production, environmental impact and mechanical properties of glubam. *Journal of Construction and Building Materials*, 44: 765–773.

Modern Engineered Bamboo Structures – Xiao, Li & Liu (eds)
© 2020 Taylor & Francis Group, London, ISBN 978-1-138-35185-1

Behavior of screwed connections in stainless steel frame shear walls sheathed with ply-bamboo panels

T. Li & Z. Li
Nanjing Tech University, Nanjing, Jiangsu, China

ABSTRACT: The behavior of sheeting connections is important in thin-walled structures, especially when diaphragm skin action is considered for these structures. This paper presents an experimental study on the monotonic and cyclic loading behavior of self-drilling screw connections in stainless steel frame shear walls sheathed with ply-bamboo panels. A total of 46 screwed connection specimens were tested to determine the effects of loading rate, end distance of screws and screw features. The results observed in this study have presented the failure modes and the effects of different factors. It is shown that, the shear behavior of screw connections is little affected by loading rate; both strength and deformation increase with the end distance increasing in the tested range; the shear behavior of phosphating self-drilling screws is superior to stainless steel self-drilling screws.

1 INTRODUCTION

Due to the advantages of good appearance, good corrosion resistance, easy maintenance and low life-cycle cost, stainless steel structure has wide applicability in building structures (Burgan. B. A et al., 2000). From the beginning of the 20th century, stainless steel has been used in building structures, initial stage was mainly used in decoration engineering, then gradually applied to building envelops and roof structures (Baddoo. N.R, 2008), but compared with traditional carbon steel, the initial cost of stainless steel structures is more higher, and the relevant design specification is not fully mature (Gardner. L, 2005). In the past 20 years, with the revision of relevant design specifications in Europe and the United States, the application of stainless-steel structures has been increasing.

Recent years, a new type of glued laminated bamboo, Glubam, has been invented, and a detailed data system about Glubam has been established (Y. Xiao et al., 2007). Meanwhile, the feasibility of ply-bamboo sheathing panels in lightweight wood frame shear walls (Y. Xiao et al., 2015) and lightweight glubam frame shear walls (R. Wang et al., 2017) has been investigated.

The main objectives of this study are to investigate the monotonic and cyclic behavior of screwed connections between stainless steel studs and ply-bamboo panels, and provide the basis for the study of the seismic behavior of framing shear walls.

2 TEST SPECIMENS

In order to study the behavior of self-drilling screw connections in stainless steel studs sheathed with ply-bamboo panels, a total of 46 screwed connection specimens were tested monotonically and cyclically. The specimen configurations listed in Table 1 differ based on loading rate, end distance of screws and screw features.

As shown in Figure 1, specimens were adopted the single-shear form, and each specimen was composed of a stainless steel stud with C-shaped section, a piece of ply-bamboo panel

Table 1. Specimen test matrix.

Test series*	Screw diameter (mm)	Screw type	End distance of screws (mm)	Loading protocol	Loading rate ($mm \cdot min^{-1}$)
SD-3.5P15-M15	3.5	P	15	M	15
SD-3.5S15-M15	3.5	S	15	M	15
SD-4.2S10-M15	4.2	S	10	M	15
SD-4.2S15-M2.5	4.2	S	15	M	2.5
SD-4.2S15-M15	4.2	S	15	M	15
SD-4.2S15-M30	4.2	S	15	M	30
SD-4.2S30-M15	4.2	S	30	M	15
SD-3.5P15-C15	3.5	P	15	C	15
SD-3.5S15-C15	3.5	S	15	C	15
SD-4.2S10-C15	4.2	S	10	C	15
SD-4.2S15-C15	4.2	S	15	C	15
SD-4.2S30-C15	4.2	S	30	C	15

*Test series name designation: SD means stainless steel studs sheathed with ply-bamboo panels; P and S represent phosphating self-drilling screw and stainless steel self-drilling screw, respectively; M and C represent monotonic and cyclic, respectively.

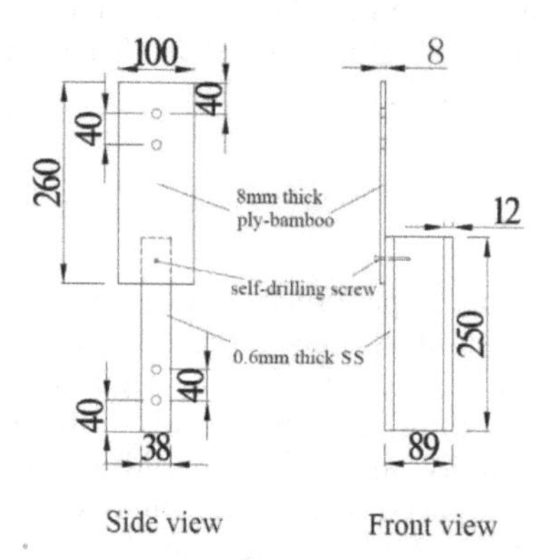

Figure 1. Diagram of screwed connections.

and a self-drilling screw. The specimens were made of the same stainless steel, which is 0.6 mm thick. The average nominal yield strength, tensile strength and elongation ratio of stainless steel according to material tests are 180 MPa, 485 MPa and 40%, respectively. The self-drilling screws used in this research are ST4.2 and ST3.5 levels circular head stainless steel screws and PT3.5 level flat head phosphating screws. Ply-bamboo panels used in this research is cross laminated ply-bamboo, the tensile strength, elastic modulus and moisture content of which are 80.59 Mpa, 9860 MPa and 7.28%, respectively.

3 EXPERIMENTAL PROGRAM

3.1 *Test setup and loading program*

As shown in Figure 2, the tests were carried out by using the 5 kN MTS electronic universal testing machine. The displacement of the specimen was measured by an LVDT (linearly variable differential transducer) with a range of ±30 mm. During the test, the load was controlled by computer according to the preset loading system. The test data were collected and recorded automatically by TST3827E dynamic and static signal test and analysis system.

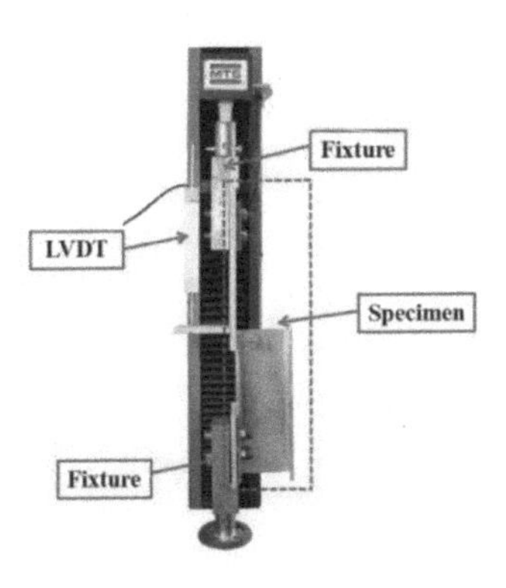

Figure 2. Test setup for connection tests.

The load was controlled by displacement with uniform speed. For cyclic loading protocol, 5 stages single cyclic loading was applied until specimen yielded. After specimen yielded, the loading step is increased by 0.5 times of yield displacement and each stage was looped three times until specimen failure.

3.2 *Experimental observations*

With the increase of load, the extrusion deformation of self-drilling screws and ply-bamboo panels gradually increased. The screw heads were gradually inclined and embedded into ply-bamboo panels, accompanied by tearing of the end of ply-bamboo panels. When the load was close to the maximum value, the stainless steel studs were gradually squeezed and warped at the contact position between screws and stainless steel studs.

As shown in Figures 3 to 4, the failure modes of screw connections mainly included tearing failure of the end of ply-bamboo panels and oblique pulled out failure of screws (pulled out of ply-bamboo panels or sliding stainless steel studs). The failure modes of specimens were mainly related to the end distance of screws, which generally presented the combination of above failure modes, and screws were well preserved after specimens failure.

3.3 *Load and displacement relationships*

Figures 5 to 9 exhibit the cyclic and corresponding monotonic load and displacement relationships of screw connection specimens. A linear elastic response can be seen from the load and

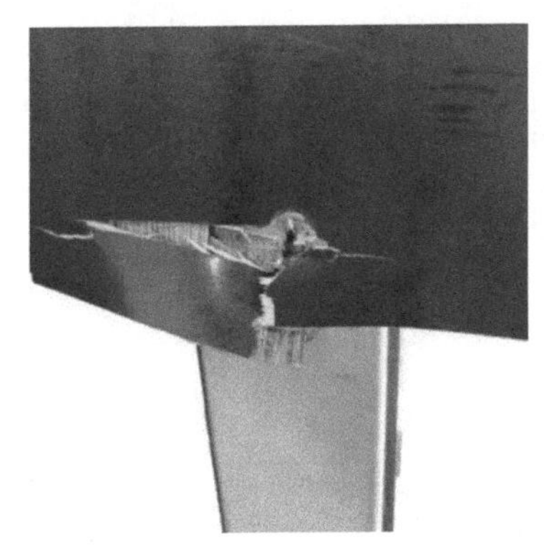

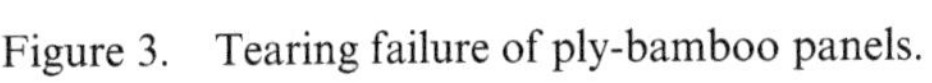

Figure 3. Tearing failure of ply-bamboo panels.

Figure 4. Oblique pulled out failure of screws.

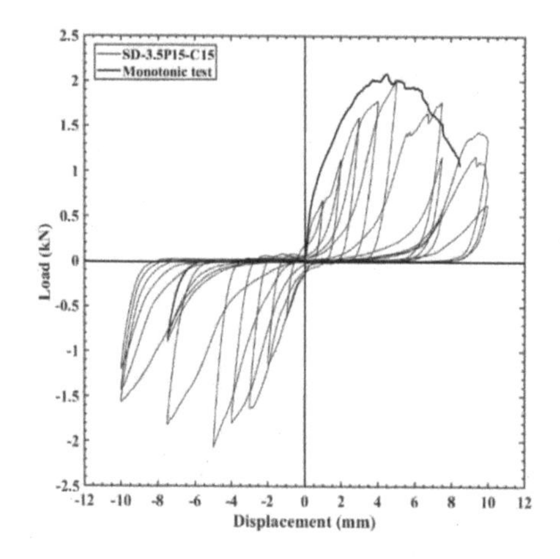

Figure 5. Hysteresis curve and corresponding monotonic curve of SD-3.5P15-C15.

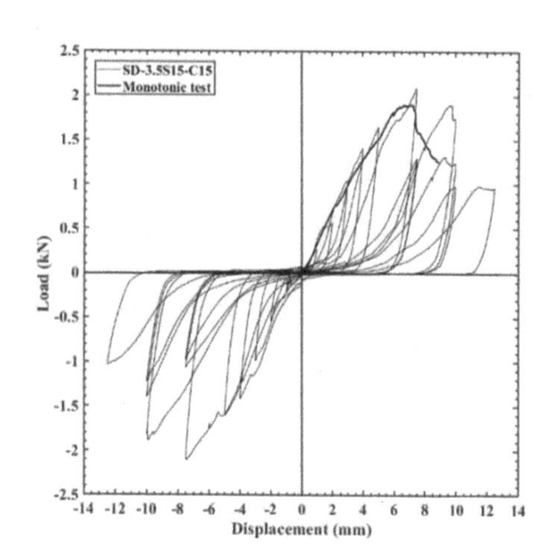

Figure 6. Hysteresis curve and corresponding monotonic curve of SD-3.5S15-C15.

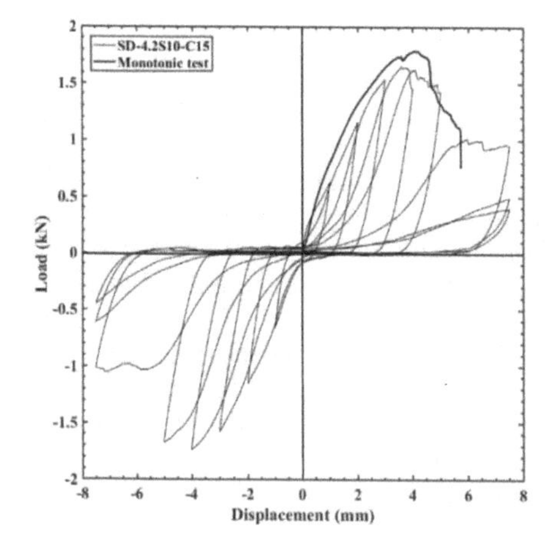

Figure 7. Hysteresis curve and corresponding monotonic curve of SD-4.2S10-C15.

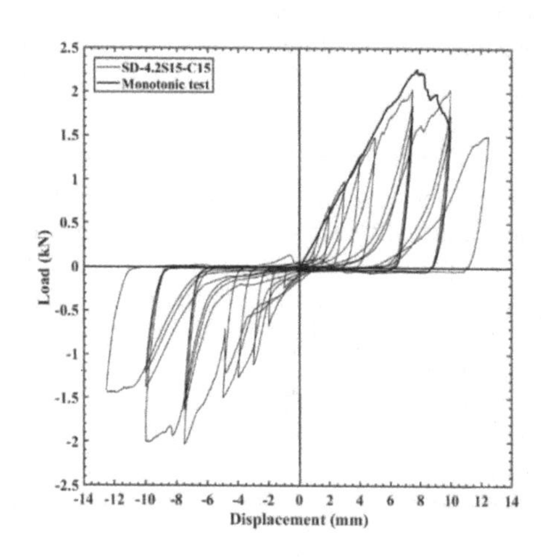

Figure 8. Hysteresis curve and corresponding monotonic curve of SD-4.2S15-C15.

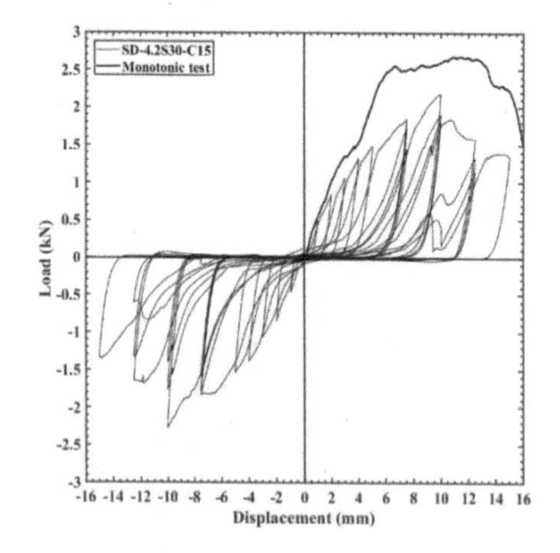

Figure 9. Hysteresis curve and corresponding monotonic curve of SD-4.2S30-C15.

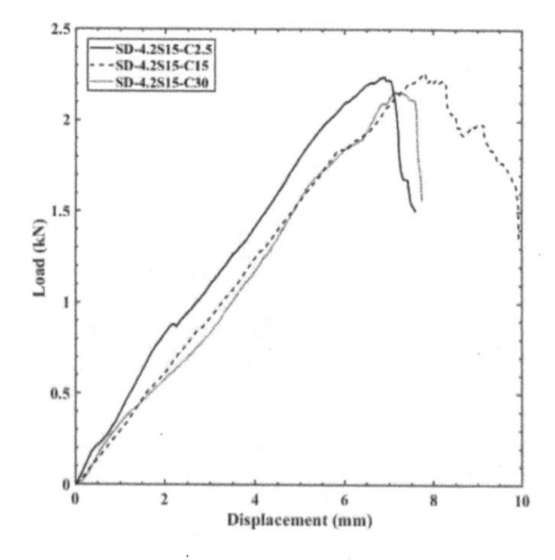

Figure 10. Load-displacement mean curves at different loading rates.

displacement relationships of specimens at the elastic stage. After this stage, the stiffness of specimens gradually reduced slightly until reaching there peak loads. Then most specimens had a distinct drop section until failure.

By and large, peak loads of most specimens in cyclic tests are lower than those recorded in monotonic tests with the average ratio of cyclic strength to static strength of 0.90. For most cyclic tests, hysteresis curves exhibit severe pinching, and strength and stiffness degrade to a certain extent corresponding to the increase of vertical displacement.

4 DISCUSSION

4.1 *Loading rate*

As comparison results shown in Figure 10, monotonic loading tests of self-drilling screw connections with three loading rates were carried out. It can be clearly seen that with the increase of loading rate, elastic stiffness presents a obvious downward trend, and the ratio of them is 1: 0.800: 0.726 (2.5 mm·min^{-1} as the baseline). Peak load has little difference between three loading rates. The ultimate displacement and ductility coefficient have no obvious change rule, and the differences among them are very small.

4.2 *End distance of screws*

Table 2 shows the different characteristic parameters of screw connection specimens at three different end distances. It can be seen that, with the increase of screw end distance from 10 mm to 15 mm, elastic stiffness declines by 50%. However, with the increase of screw end distance from 15 mm to 30 mm, elastic stiffness has no obvious change. Besides, peak load, peak displacement and ultimate displacement increase with the increase of screw end distance, but ductility coefficient has no obvious change rule.

4.3 *Screw features*

In this research, two types of self-drilling screws, phosphating self-drilling screw and stainless steel self-drilling screw, are adopted, that the diameter of phosphating self-drilling screw is 3.5 mm, while the diameters of stainless steel self-drilling screw are 3.5 mm and 4.2 mm, respectively.

As can be seen from Table 3, elastic stiffness, peak load and ductility coefficient of phosphating self-drilling screw connection specimens are higher than that of stainless steel self-drilling screw connection specimens to varying degrees. Among them, ductility coefficient is the biggest difference. Therefore, when using self-drilling screws of the same diameter, phosphating self-drilling screws are superior to stainless steel self-drilling screws. In addition, parameters have no obvious fluctuation with the screw diameter increasing from 3.5 mm to 4.2 mm.

Table 2. Comparison of characteristic parameters for specimens at different end distances.

Test series	Elastic stiffness (N·mm^{-1})	Displacement at peak load (mm)	Peak load (N)	Ultimate displacement (mm)	Ductility coefficient
SD-4.2S10-M15	841	4.19	1783	4.65	1.989
SD-4.2S15-M15	308	8.06	2223	9.20	1.628
SD-4.2S30-M15	501	11.67	2661	14.63	2.873
SD-4.2S10-C15	562	3.94	1599	4.98	2.196
SD-4.2S15-C15	276	7.49	1980	9.97	1.760
SD-4.2S30-C15	294	9.96	2174	10.90	1.612

Table 3. Comparison of characteristic parameters for specimens with different screws.

Test series	Elastic stiffness ($N \cdot mm^{-1}$)	Displacement at peak load (mm)	Peak load (N)	Ultimate displacement (mm)	Ductility coefficient
SD-3.5P15-M15	591	4.59	2206	6.40	2.214
SD-3.5S15-M15	351	7.38	1833	7.77	1.398
SD-4.2S15-M15	308	8.06	2223	9.20	1.628
SD-3.5P15-C15	453	4.98	1986	7.44	1.962
SD-3.5S15-C15	293	7.45	2070	9.99	1.691
SD-4.2S15-C15	276	7.49	1980	9.97	1.760

5 CONCLUSIONS

A total of 46 stainless steel studs sheathed with ply-bamboo panels were tested monotonically and cyclically to analyse the effect on screw shear behavior of three factors. Firstly, the shear behavior of screw connections is little affected with the increase of loading rate. Secondly, with the increase of screw end distance from 10 mm to 15 mm, elastic stiffness presents a obvious downward trend. Besides, peak load, peak displacement and ultimate displacement increase with the increase of screw end distance. Thirdly, with the same diameter, phosphating self-drilling screws are superior to stainless steel self-drilling screws, and shear behavior has no obvious fluctuation with the screw diameter increasing from 3.5 mm to 4.2 mm. The results obtained in this study can further promote the application of bamboo-based material in stainless structures.

ACKNOWLEDGEMENT

The research described in this paper was sponsored by the Institute of Advanced Bamboo Based Materials and Structures (ABMS) at Nanjing Tech University, under the support of National Science Foundation of China (51678302).

REFERENCES

Burgan, B.A., Baddoo, N.R. & Gilsenan. K.A. 2000. Structural design of stainless steel members-comparison between Eurocode 3, Part 1, 4 and test results. *Journal of Constructional Steel Research* 54 (1): 51–73.

Baddoo, N.R. 2008. Stainless steel in construction: A review of research, applications, challenges and opportunities. *Journal of Constructional Steel Research* 64 (11): 1199–1206.

Gardner, L. 2005. The use of stainless steel in structures. Progress in Structural Engineering and Materials 7 (2): 45–55.

Xiao, Y., Shan, B., Chen, G., Zhou, Q. & She, L.Y. 2007. Development of a new type Glulam-Glubam, *Proceedings of the First International Conference on Modern Bamboo structures (ICBS-2007); Changsha, 28-30 October 2007.*

Xiao, Y., Li, Z., Wang R. 2015. Lateral loading behaviors of lightweight wood-frame shear walls with ply-bamboo sheathing panels. *ASCE Journal of Structural Engineering* 141 (3): B4014004.

Wang, R., Xiao, Y., Li, Z. Lateral Loading Performance of Lightweight Glubam Shear Walls. *ASCE Journal of Structural Engineering* 143 (6): 04017020.

Modern Engineered Bamboo Structures – Xiao, Li & Liu (eds)
© 2020 Taylor & Francis Group, London, ISBN 978-1-138-35185-1

Experimental study on axial compression of reconsolidated bamboo columns

Y.Q. Liu & B.W. Chen
Central South University of Forestry and Technology, Changsha, Hunan, China

ABSTRACT: Reconsolidated bamboo is a kind of biological materials that reorganizes and strengthens bamboo materials. This article has discussed the short term compression test of the reconsolidated bamboo column. Short-term test is mainly about the axial compression mechanical properties of the center column and the medium long column. We have researched 5 groups of different slenderness ratio of reconsolidated bamboo columns, including the vertical strain, lateral strain, lateral displacement and the rule about the ultimate bearing capacity changing with the slenderness ratio, then discussed failure pattern and failure mechanism of different slenderness ratio columns. The test results show that the reconsolidated bamboo columns has good mechanical properties, and the failure mode gradually transfer into the buckling failure from the strength failure.

1 INTRODUCTION

Along with the economic and social development, as well as the increasingly strengthening environmental protection consciousness, the building materials with bamboo and wood as raw materials have been more and more widely used. By taking natural bamboo as the raw material and using advanced recombination technology, the bamboo wood can be made into building materials of different sizes through drying, gluing, hot press molding and other processes, so as to meet the new requirements in the civil engineering field. The modern engineering bamboo represented by reconsolidated bamboo has appeared on the historical stage, which has many advantages, such as high strength, compact material, and highly industrialized production mode and etc.

As a new type of building material, the current research on reconsolidated bamboo timber mainly focuses on processing, market study, basic performance and etc. Nevertheless, there are few related studies on its structural components of building structures, and there is no relevant architectural design standard for reconsolidated bamboo at home and abroad currently. Therefore, the axial compression test was carried out on the reconsolidated bamboo column in this paper, including short-term compression, and the failure performance and mechanism was also studied experimentally.

2 SHORT-TERM TEST

2.1 *Production of specimens*

The specimen was composed of 4-layer reconsolidated bamboo boards, with a section of 100*100 mm, as shown in Figure 1. A total of 15 specimens were made, which were divided into 5 groups according to the slenderness ratio, with the lengths of 600 mm, 800 mm, 1,000 mm, 1,400 mm and 1,600 mm respectively.

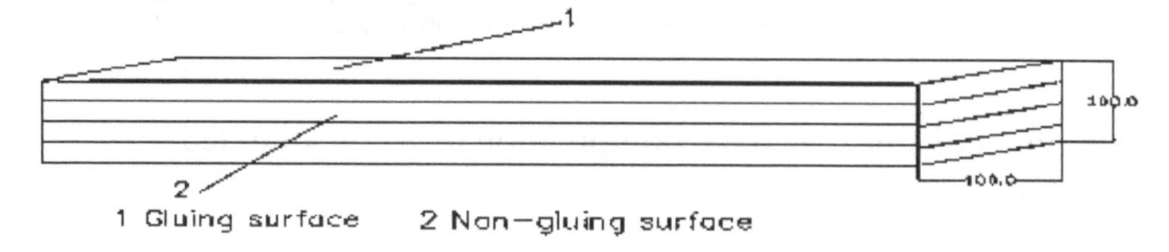

Figure 1. Specimen.

2.2 *Test equipment and test points arrangement*

The test was completed by a 5000 kN microcontroller electro-hydraulic servo press from the Central South University of Forestry and Technology, which could collect the load on specimens and the vertical displacement. The strain collection was conducted by TST static strain testing system, and the bidirectional knife hinge was used as the support. In order to measure the displacement and deformation of specimens during the test, the vertical and horizontal strain gauges were arranged on the central four sides of the column, at 1/2 of the specimen in length direction, the displacement meters were respectively arranged on two adjacent surfaces to measure lateral displacement. The loading device and measuring points were arranged as shown in Figure 2.

In order to ensure the axial compression of specimens, the combination method of physical alignment and geometric alignment was adopted in this test. When installing specimens, it was necessary to ensure that, the middle line of specimen section, the center of bidirectional knife hinge, and the center of press were on the same axis, so as to ensure geometric alignment; before formally loaded in the test, the specimen should be preloaded, and the vertical strain on four sides of the reconsolidated bamboo column should be observed during preloading. The vertical strain reading difference of the four sides should be ensured to be less than 5%; otherwise, it was required to adjust the specimen before preloading, so as to ensure physical alignment. It was loaded to failure at a constant speed of 1.5 mm/min in the test, and the strain and displacement readings were collected by the static strain gauge at a frequency of 5 s/time.

2.3 *Test results and analysis*

The bearing capacity and failure morphological changes of the 5 groups of tests were statistically analyzed, as shown in Table 1. In the table, N_{max} indicated the maximum bearing capacity, and Δ was the reduction value of the axial peak load of the specimen relative to column A.

Figure 2. The loading device and measuring points.

Table 1. Statistics of the characteristic parameters for the reconsolidated bamboo columns.

Specimen number	l_0/(mm)	λ	N_{max}/ (kN)	Δ/(%)	Failure mode
A-1—A-3	600	20.8	686.4	0	Strength fracture
B-1—B-3	800	27.7	676.6	1.6	Strength fracture
C-1—C-3	1000	34.7	582.8	15.1	Strength fracture
D-1—D-3	1400	48.5	520.4	24.2	Buckling failure
E-1—E-3	1600	55.4	440.7	35.8	Buckling failure

2.3.1 *Failure characteristics*

Figure 3 (A-1) and Figure 3 (C-1) show the strength fracture. When the ultimate load was reached, there was a sound of rubber layer off, cracks appeared on the top or bottom of the specimen, the load began decreasing, and the crack continued to extend. Figure 3 (E-1) shows the buckling failure, when the ultimate load was reached, cracks appeared firstly on the tensile side of specimen, and the bamboo fibers were ruptured and accompanied by slight sound; at this point, the load began decreasing, and cracks began to expand around, thereby resulting in buckling failure due to excessive transverse deformation. In the loading process of all specimens, it was an elastic state in an early stage, and it had obvious deformation and good deformation recovery ability in a later stage, which was of great significance to the application of reconsolidated bamboo in the structure.

2.3.2 *Load-strain relationship*

According to the load-strain relationship of three groups of specimens given in Figures 4 to 6, the following conclusions can be drawn.

The vertical strain and transverse strain change trend of each group of specimens were basically the same, among which the elastic segment of transverse strain was longer. Both could be divided into the elastic working stage, plastic working stage, and failure stage. At the early stage of loading, the load and strain showed a linear relationship, as the load increased, and changed, the curve slope decreased, and it transformed into a plastic working stage, after it reached the peak load of the specimen, the strain of the specimen also reached the peak, and the specimen lost its bearing capacity.

The peak strain of reconsolidated bamboo column decreased with the increase of slenderness ratio, and the peak strain of B to E column decreased by 9.8%, 13.2%, 18.6% and 28.1% respectively compared with that of A column. When it reached about 90% of the ultimate load, the lateral strain of each group had a large change in slope, indicating that the deformation of reconsolidated bamboo column began to increase sharply at this time, thus resulting in relatively obvious bending deformation.

2.3.3 *Load-displacement relationship*

The load-displacement curve of each group was obtained according to the experimental data, as shown in Figures 7 to 9. The following conclusions can be drawn from the analysis of data in the figures:

It can be seen from Figure 7 and Figure 8 that, each group of specimens showed a tiny displacement when loaded at the beginning, which was mainly due to the initial imperfection of reconsolidated bamboo column, in the early stage of loading, the lateral displacement and load of each specimen showed a linear relationship, and the column was mainly subject to compressive stress at this point, when the load reached about 85% of the ultimate load, the lateral displacement began increasing suddenly, and the curve slope decreased, after it reached the limit load, the lateral displacement increased continuously.

It could be seen in Figure 9 that, the lateral displacement and load of specimens showed a linear change, however, its linear phase decreased significantly with the increase in slenderness ratio, and when about 55% of the ultimate load was reached, the lateral displacement

A-1 C-1 E-1 Test site

Figure 3. Failure mode of reconsolidated bamboo column.

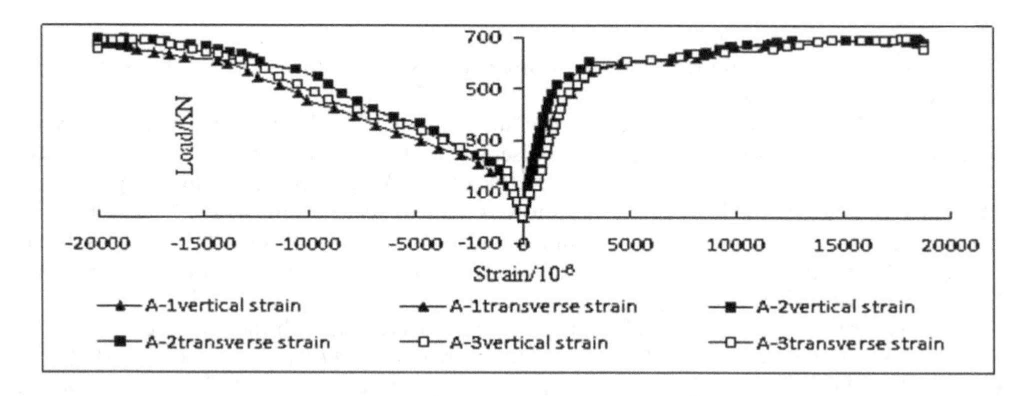

Figure 4. Load -strain curve of column A.

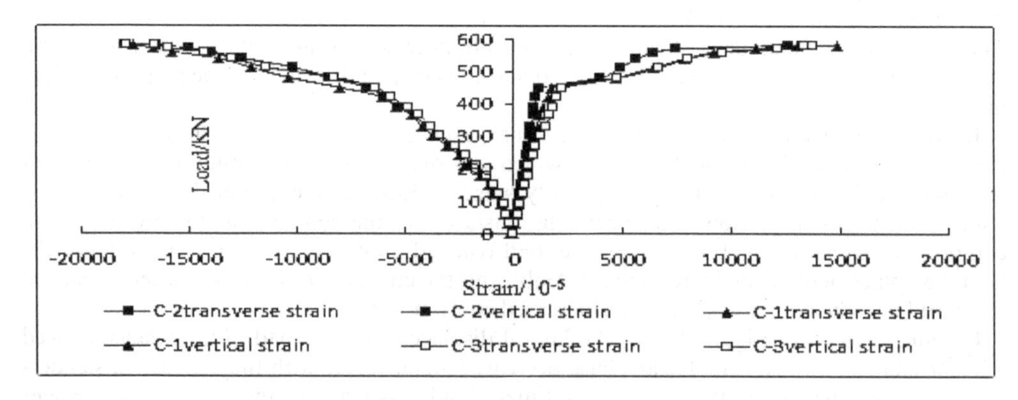

Figure 5. Load -strain curve of column C.

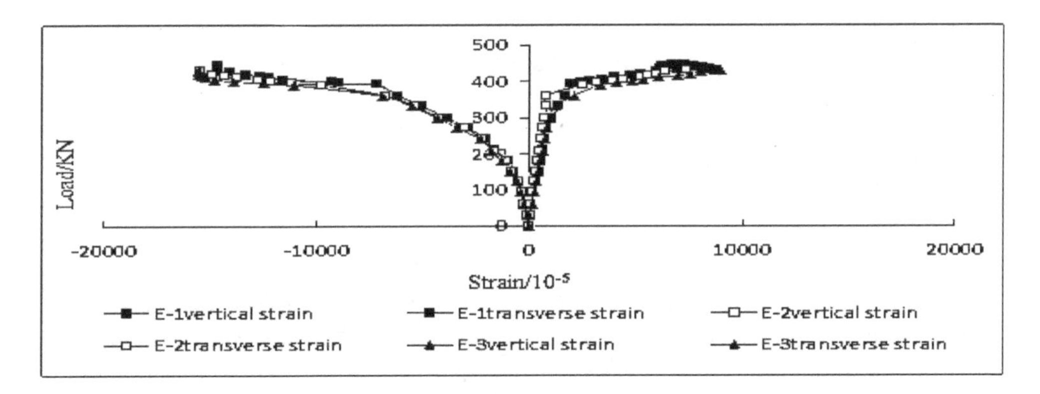

Figure 6. Load -strain curve of column E.

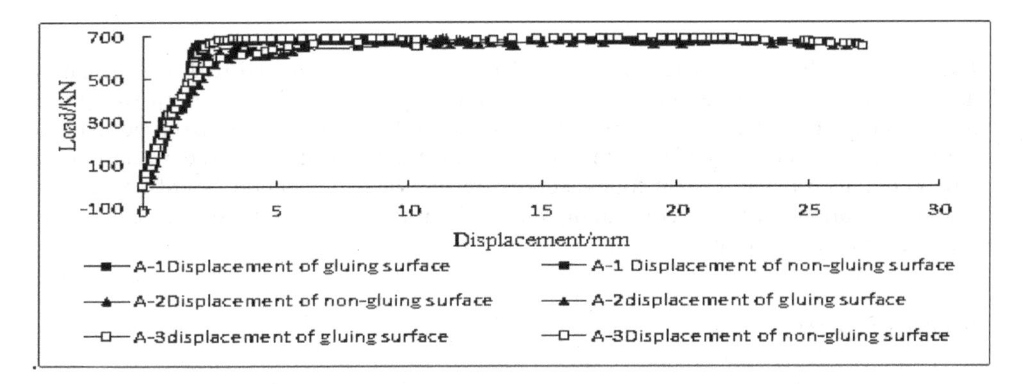

Figure 7. Curves of load-displacement for column A.

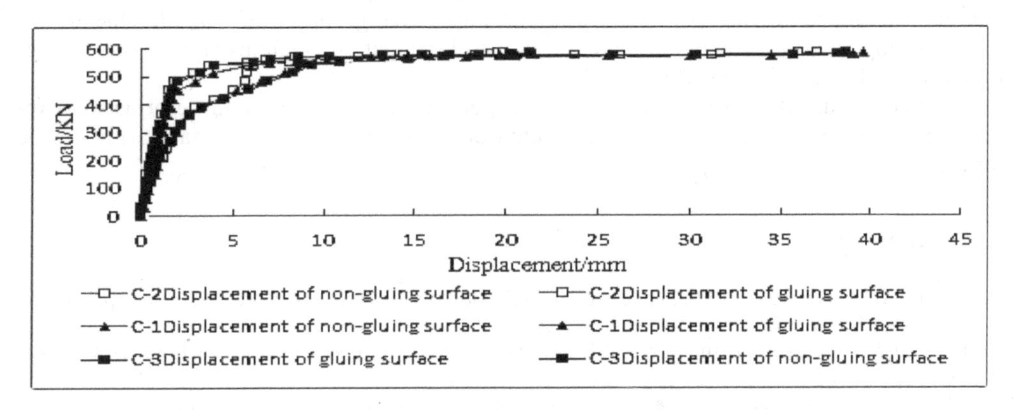

Figure 8. Curves of load-displacement for column C.

began to increase obviously, and the obvious buckling failure appeared, simultaneously, the displacement of gluing surface and non-gluing surface of column E was very obvious, and the crack area was in the binding site of gluing surface and non-gluing surface in combination with the failure phenomenon of column E shown in Figure 3, which showed that the long column was prone to bi-directional buckling failure.

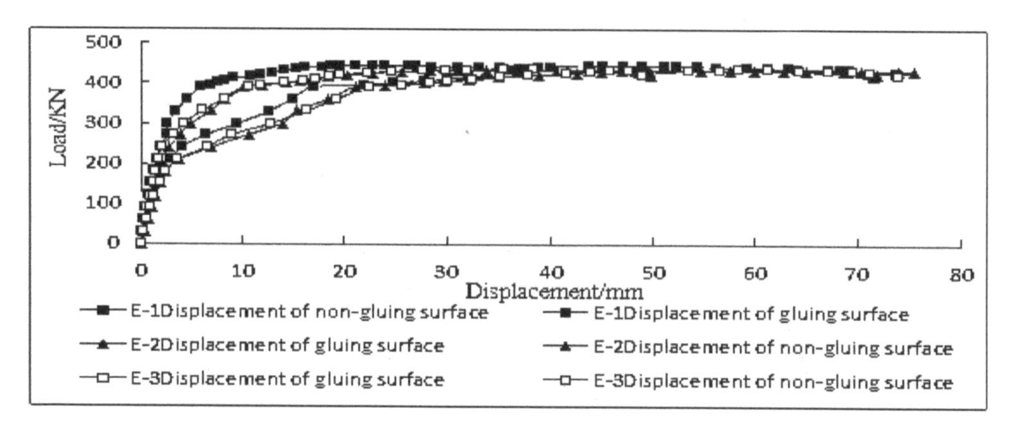

Figure 9. Curves of load-displacement for column E.

The lateral displacement of reconsolidated bamboo column increased with the increase of slenderness ratio. The displacement on the non-bonding surface of each group was greater than that on the bonding surface, which was caused by the anisotropy of bamboo material, and the different bending stiffness in different directions. Meanwhile, the reconsolidated bamboo column was formed by four layers of reconsolidated bamboo boards, which had a certain lamination effect, thereby making the bending stiffness of gluing surface greater than that of the non-gluing surface.

3 CONCLUSIONS

Under the action of short-term load, the reconsolidated bamboo column gradually changed from strength failure to buckling failure with the increase of slenderness ratio, and its peak load and peak strain gradually decreased.

With the increase in slenderness ratio, the lateral displacement of the reconsolidated bamboo column gradually increased, and the linear change phase decreased. Due to the material properties of the bamboo material, the displacement of the gluing surface was smaller than that of the non-gluing surface.

The reconsolidated bamboo column showed good bearing capacity and deformation capacity in the test, mainly manifested to be a kind of good construction material during ductile failure.

REFERENCES

Bengtsson, M., Oksman, K. 2006. Silane crosslinked wood plastic composites: Processingand properties. *Composites Science and Technology* 66(13): 2177–2186.

Li, H.T. 2015. Experimental study on PSBL under eccentric compression. *Journal of Building Materials* 19(3): 561–565.

Su, J.W. 2015. Experimental research on parallel bamboo strand lumber column under axial compression. *China Science Paper* 10(1): 48–51.

Sun, L.L. 2013. Research on stress-strain relationship in parallel-to-graindirection of parallel strand bamboo. *Nanjing Forestry University*.

Wu, P.Z. 2015. Experiment and analysis on creep properties of reconsolidated bamboo. *Northeast Forestry University*.

Xiao, Y. 2015. Experimental studies of glubam columns under axial loads. *Industrial Construction* 6(4): 13–17.

Zhang, K.B. 2011. Experimental research on creep of reinforced concrete pillars. *Journal of Changsha University of Science and Technology* 8(2): 17–21.

Modern Engineered Bamboo Structures – Xiao, Li & Liu (eds)
© 2020 Taylor & Francis Group, London, ISBN 978-1-138-35185-1

Analysis of elastic-plastic bending of prefabricated composite beams considering slip effect

J.W. He, S.W. Duan & J.R. Wang
Central South University of Forestry and Technology, Changsha, Hunan, China

ABSTRACT: The prefabricated composite steel-wood beams will create slip effect on the interface. It will increase deformation, reduce beams' bearing capacity and influence specimens' use function. In order to analyze composite beams' mechanical properties considering slip effect on the interface, eight composite beams have been fabricated and bending test have been finished. Hot-Rolled H steel is used as the main framework, and the larch is fixed on the two flanges of steel by two ways–bolts and epoxy resin glue. On the basis of the elastic-plastic assumption, calculation models of composite beams have been established. They are mainly about integral bending moment, local bending moment, section stress and yield bending moment. Meanwhile, the calculation formulas of section elastic stress and yield bending have been given. Compared with test results, yield moment calculated from models on the section is smaller than the theory, and it has better accuracy. Therefore, it provides a convenient and feasible method to calculate beam section stress and yield bending moment.

1 INTRODUCTION

The prefabricated composite steel-wood beam is a new kind of composite component. It is applied to composition materials of framed buildings by using "steel" and "wood", which are combined by high-strength big hexagon bolts and epoxy resin adhesive (Chen, W., 2016; Shi, Y., 2006; Wang, Z.H., et al., 2011; Cao, L., 2017; Yu, J., 2004). Hot-Rolled H steel is used as the main framework of buildings; wood is used to work as concrete. It is fastened on the surface of the upper and lower flange from steel. On one hand, wood can prevent steel from rusting by touching less air, on the other hand, wood is also used to bear load with steel. Therefore, the character of energy-saving materials from wood can be applied, and it accords closely with the environmental-friendly demands in our contemporary society (Qiu, B.X., 2005; Liu, C.F., et al., 2015; Sakuragawa. S., et al., 2005; Liu, G.Z., 2012).

Hot-Rolled H steel and larch having been processed are used as the main composite component of composite beams. According to the code for design of timber structures GBT 50005-2003, epoxy resin adhesive is used to slice between the surface of larch and upper and lower flange from steel, and big hexagon bolts are used to further fasten at every equidistant position of the composite beams (Lu, W.D., et al.,2011; Li, H.B., 2010; Gao, H.J., et al., 2002; Gao, L.Y., et al., 2006). In this study, the elastic-plastic analysis methods to composite beams are transformation-section method, which converts the larch into steel by rigidity (Jiang, J.J., 1998; Johnson. R. P, 2004; Nie, J.G., et al., 2005), strain and so on. The section is analyzed as the single and successive materials.

For the composite beams, if there is no slip effect between steel and larch, the section stress should be calculated basing on whole beams. Some shear connection exists in composite beams during the process of building, and slip effect exists on the section. It has been researched as one of the point contents due to its influence to composite beams' strength, rigidity and computation methods (Johnson. R P., et al., 1991; Nie, J.G., et al., 1995; Nie, J.G., et al., 1997; Yin, Y.Q., et al., 1993; Jiang, X.G., et al., 2007). Professor Nie Jianguo proposed the computation methods about elastic stress and strain on section considering slip

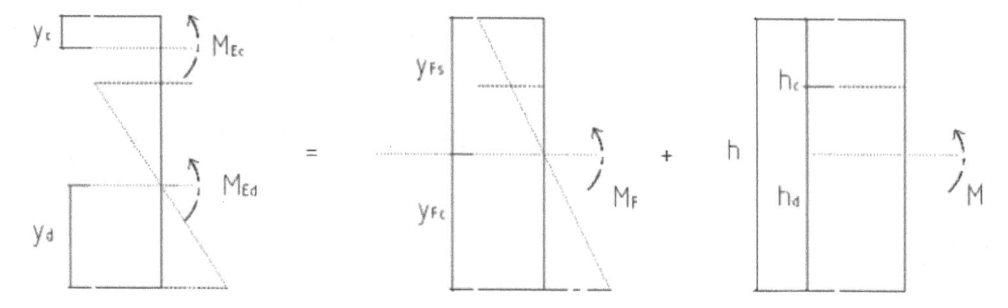

Figure 1. Bending moment combination of composite beam section.

effect. Some part of his achievements have been adopted by many domestic scholars when they analyse in related fields.

Basing on the character of slip effect, we have changed compatibility condition, physical condition and balance condition. According to those conditions,computation formula about integral bending moment and local bending moment for composite beams have been established, and computation methods to elastic stress on section have also been gained (Jiang, X.G., et al., 2007).

2 BASE ASSUMPTION

a. Steel and wood are linear elastic materials.
b. Bending curvature from larch and steel is always equal without considering lift effect.
c. Equivalent stiffness method is to be applied to convert section, and transformation-section will be expressed by properties of steel.
d. There is no consideration to steel' s holistic instability and local buckling.
e. There is slip effect between larch and steel, and the computation methods of slip strain ε_s is confirmed by conversion stiffness methods which were proposed by professor Nie Jianguo.
f. The total bending moment M on section is composed by two parts: integral bending moment M_F acts on composite beams, it will not introduce section slip; local bending moment M_E acts on laminated beams, and wood will bear bending moment M_{Ec}; steel will bear bending momentM_{Ed}, and slip effect will be produced on section. That is $M = M_F + M_E$, $M_E = M_{Ec} + M_{Ed}$, and bending moment on section just can be seen from Figure 1.

3 LOCAL BENDING MOMENT

Being acted by local bending moment, bending curvature of larch and steel is equal, and the formula can be expressed as follows:

$$\frac{M_{Ec}}{EI_c/\alpha_E} = \frac{M_{Ed}}{EI_d} \tag{1}$$

EI_c/α_E and EI_d are bending stiffness from larch and steel respectively, whereα_E is elastic modulus ratio of steel and larch.

Being acted by local bending moment, the sum of larch' s tensile strain and steel upper flange's compressive strain on section are slip effect. The formula can be expressed as follows:

$$\varepsilon_{Ec} + \varepsilon_{Ed} = \varepsilon_s \tag{2}$$

Thereinto, the edge strain from larch is:

$$\varepsilon_{Ec} = \frac{M_{Ec}}{EI_c/\alpha_E} y_c \tag{3}$$

The strain from steel upper flange is:

$$\varepsilon_{Ed} = \frac{M_{Ed}}{EI_d} y_d \tag{4}$$

The slip effect is:

$$\varepsilon_s = h\Delta\phi = h\frac{\zeta M}{EI} \tag{5}$$

Putting Equation (3) ~ (5) into Equation (2), the formula can be expressed as follows:

$$\frac{M_{Ec}}{EI_c/\alpha_E} y_c + \frac{M_{Ed}}{EI_d} y_d = \frac{\zeta M h}{EI} \tag{6}$$

In above-mentioned formulas, y_c is distance from centroid in larch to interface, y_d is distance from centroid in steel to interface, h is the total height of composite section, $\Delta\phi$ is the section curvature produced by interface slip effect, section ζ is rigidity reduced factor, and EI is bending stiffness of calculation on section from composite beams.

Putting equation (1) into equation (6), the formula can be expressed as follows:

$$\frac{M_{Ed}}{EI_d} y_c + \frac{M_{Ed}}{EI_d} y_d = \frac{\zeta M h}{EI}$$

We can simply transform it as follows:

$$\frac{M_{Ed}}{EI_d}(y_c + y_d) = \frac{\zeta M h}{EI}, \text{ that is } \frac{M_{Ed}}{EI_d} d_0 = \frac{\zeta M h}{EI}$$

In this formula, $d_0 = y_c + y_d$ is the centroid distance of composite beams.

The local bending moment of steel is as follows:

$$M_{Ed} = \frac{EI_d}{EI} \cdot \frac{h}{d_0} \cdot \zeta M = \frac{hI_d}{d_0 I} \zeta M \tag{7}$$

The local bending moment of larch is as follows:

$$M_{Ec} = \frac{EI_c/\alpha_E}{EI} \cdot \frac{h}{d_0} \cdot \zeta M = \frac{hI_c}{\alpha_E d_0 I} \zeta M \tag{8}$$

4 INTEGRAL BENDING MOMENT

According to the constituent of section bending moment, the integral bending moment of section can be got as follows:

$$M_F = M - M_E = M - M_{Ec} - M_{Ed} = M - \frac{EI_d}{EI} \cdot \frac{h}{d_0} \cdot \zeta M - \frac{EI_c/\alpha_E}{EI} \cdot \frac{h}{d_0} \cdot \zeta M$$
$$= \left(1 - \zeta \frac{EI_0}{EI} \frac{h}{d_0}\right) M \tag{9}$$

There into: $EI_0 = EI_c/\alpha_E + EI_d$
Order:

$$M_E = \varsigma M \tag{10}$$

from Equation (9), the coefficient ς of integral bending moment is as follows:

$$\varsigma = 1 - \zeta \frac{hI_0}{d_0 I} \tag{11}$$

5 SECTION STRESS

When slip effect is considered, section stress can be obtained by superposition principle from theory of material mechanics:
steel stress:

$$\sigma_s = \frac{M_F}{I} y_{Fs} + \frac{M_{Ed}}{I_d} y_d \tag{12}$$

the maximum value of larch is:

$$\sigma_c = \frac{M_F}{\alpha_E I} y_{Fc} + \frac{M_{Ed}}{I_c} y_c \tag{13}$$

From this formula, I is moment of inertia on composite section, $I_c$is inertia moment of larch, I_d is inertia moment of steel, y_{Fs} is distance from lower flange of steel to centroid in composite beams section, y_d is distance from lower flange of steel to centroid in steel, y_{Fc} is distance from larch's upper edge to centroid in composite beams section, and y_c is distance from larch's upper edge to centroid in larch.

6 YIELD BENDING MOMENT

Putting Equation (9) and Equation (7) into Equation (13), elastic stress of steel lower flange about composite section can be expressed as follows:

$$\sigma_s = \frac{M_F}{I} y_{Fs} + \frac{M_{Ed}}{I_d} y_d = \left(1 - \zeta \frac{hI_0}{d_0 I}\right) M \frac{y_{Fc}}{I} + \frac{(hI_d/d_0 I)\zeta M}{I_d} y_d = \left[\left(1 - \zeta \frac{hI_0}{d_0 I}\right) y_{Fc} + \zeta \frac{h}{d_0} y_d\right] \frac{M}{I} \tag{14}$$

When lower flange of steel reaches to steel yield strength and yield strength is viewed as the yield limit state of section, the yield bending moment of composite beams can be expressed as follows:

$$M_y = \frac{I}{\left(1 - \zeta \frac{hI_0}{d_0 I}\right) y_{Fc} + \zeta \frac{h}{d_0} y_d} f_y \tag{15}$$

When there is test value of slip effect from composite beams, stiffness reduction factorζcan be calculated by Equation (5):

$$\zeta_{test} = \frac{\varepsilon_{s,test} EI}{hM_{test}} \tag{16}$$

From Equation (16) and Equation (15), when slip strain is greater, section stiffness reduction factor will be greater, local bending moment and the local stress from it will be greater. The yield bending moment of composite beams section will be smaller.

For design issues, stiffness reduction factor ζ can be calculated according to arrangement of shear connectors. Its theory can be referred to (Nie, J.G., et al., 1995), and the simplified value can be referred to (Wang, P., et al., 1996).

7 COMPARATIVE ANALYSIS OF EXPERIMENT

Geometric dimensions and material parameters of specimens can be seen as Table 1. The wood beams are glued and compressed by several pieces of larch. The yield bending moment of composite beams is calculated by (15), and it can been seen from Table 2.The comparisons between yield bending moment calculation and test value from document (Xu, S.L., et al., 2001) can be seen from Table 2.

Table 1. Geometrical sizes and material parameters.

Specimens	Materials	Are there bolts?	Beams length (mm)	Beams height (mm)	Beams width (mm)	Thickness of upper and lower wood (mm)	Width of upper and lower wood (mm)	Steel (mm)
			l	h	b	$h_1 = h_2$	$b_1 = b_2$	$h_0 \times b_0$
SWL1	Steel Larch	Yes	2000	200	75	25	75	150 × 75
SWL2	Steel Larch	No	2000	200	75	25	75	150 × 75
SWL3	Steel Larch	Yes	2000	200	100	25	100	150 × 75
SWL4	Steel Larch	No	2000	200	100	25	100	150 × 75
SWL5	Steel Larch	Yes	2000	250	100	25	100	200 × 100
SWL6	Steel Larch	No	2000	250	100	25	100	200 × 100
SWL7	Steel Larch	Yes	2000	250	125	25	125	200 × 100
SWL8	Steel Larch	No	2000	250	125	25	125	200 × 100
WL1	Larch	No	2000	150	75			
WL2	Larch	No	2000	200	100			

Note: a The model of Hot-Rolled H steel is Q235; b The average value of yield strength from Q235 steel is 270.6 MPa, and it can be calculated by standard value of strength and intensity variation coefficient which is 0.08 from steel.

Table 2. Yield moment of steel-wood composite beams.

Specimens	$\varepsilon_{s,test}$ ($\times10^{-6}$)	$\zeta_{calculation}$	$M_{y,test}$ (kN·m)	$M_{y,noslip}$ (kN·m)	$M_{y,paper}$ (kN·m)	$\frac{M_{y,paper}}{M_{y,test}}$	$\frac{M_{y,document}}{M_{y,test}}$
SWL1	3156	0.309	162.8	166.2	156.7	0.963	1.029
SWL2	3276	0.312	160.3	162.1	153.5	0.956	1.121
SWL3	3301	0.328	158.4	159.6	151.1	0.954	1.065
SWL4	3590	0.359	152.7	158.9	143.8	0.942	1.098
SWL5	3720	0.346	149.8	155.6	142.7	0.952	1.102
SWL6	3810	0.378	150.3	153.8	148.5	0.988	1.007

According to Table 2, we can see as follows: When slip effect is considered, yield bending moment calculation of composite beams is close to test value, and the calculation is smaller than test value. It conforms to security of industrial design and analysis.

The yield bending moment calculated by the method from this study is smaller than the value from document (Nie, J.G., et al., 1997). The main reason is the major premise in this study, and it assumes that complete bending of composite beams is resulted from interface slip strain. With the same interface slip, curvature of section is greater than the calculation from document (Nie, J.G., et al., 1997), and the elastic stress on section from this study is greater than the calculation from document (Nie, J.G., et al., 1997).

8 CONCLUSIONS

a. According to the models acted by integral bending moment and local bending moment respectively, when slip effect is considered, a method to calculate the prefabricated composite steel-wood beams has been proposed.
b. The calculation results of composite beams from this study conforms to test value, and they fit security.
c. When slip effect is smaller, section stiffness reduction factor will be greater. The yield bending moment of composite beams is related to interface slip and yield bending moment will be smaller.

ACKNOWLEDGEMENT

The study was supported by National Science Foundation of China under Grant no. 5127425.

The experiment was supported by Hunan Engineering Laboratory of Modern Wood Structure Engineering Materials Manufacturing and Application Technology.

REFERENCES

Cao, L. 2017. The research on mechanical properties of larch glulam beams. *Central South University of Forestry & Technology.*

Chen, W. 2000. Mechanical behaviour analysis and experimental study on engineered timber-concrete composite beams. *Central South University of Forestry & Technology.*

Gao, H.J., Xi, T. & Sun, F.F. 2002. The fabrication and application of hot-rolled H-shaped steel. *Progress in Steal Building Structures* 4(1): 33–44.

Gao, L.Y., Deng, C.G. & Hao, J.P. 2006. Study on integral stability of homemade rolled H-section beams-applicability of calculation principle of stability coefficient to homemade hot-rolled H-section beams. *Industrial Construction* 36(10): 75–78.

Jiang, J.J. 1998. Concrete structure engineering. Beijing: *China Architecture and Building Press.*

Jiang, X.G., Ju, J.S. & Fu, X.R. 2007. Analysis of elastic stress of composite steel-concrete beams considering slip effect. *Engineering Mechanics* (01): 143-146+142.

Johnson R P. & Molenstra N. 1991. Partial shear connection in composite beams in buildings [A]. *Proc. Inst. Civ. Engrs* Part 2, 91: 679~704.
Johnson, R.P. 2004. Composite structure of steel and concrete. Vol 1: beams, columns, frames and applications in buildings. 3rd ED. *Oxford: Blackwell Scientific.*
Li, H.B. 2010. FEM analysis of head deformation during forming H-beam. *Hebei Polytechnic University.*
Liu, C.F. & Liu, F. 2015. Introduction to realize the approach of the sustainable development green building. *Chinese & Overseas Architecture* (3): 69–71.
Liu, G.Z. 2012. Study on the design and construction of modern wood structure residence. *Northeast Forestry University.*
Lu, W.D.,Yang, H.F., Liu, W.Q., Yue, K. & Chen, X.W. 2011. Development, application and prospects of glulam structures. *Journal of Nanjing University of Technology (Natural Science Edition)* 33(5): 79–81.
Nie, J.G. & Shen, J.M. 1997. Slip effect on strength of composite steel-concrete beams. *China Civil Engineering Journal* 30(1): 31~36.
Nie, J.G., Liu, M. & Ye, L.P. 2005. Composite structures of steel and concrete. Beijing: *China Architecture and Building Press* 59~60.
Nie, J.G., Shen, J.M. & Yu, Z.W. 1995. A reduced rigidity method for calculating deformation of composite steel-concrete beams. *China Civil Engineering Journal* 28(6): 11~17.
Qiu, B.X. 2005. Promoting green building and speeding up the construction of a resource conserving society. *China Construction News.*
Sakuragawa, S., Miyazaki, Y., Kaneko T & Makita T. 2005. Influence of wood wall panels on physiological and psychological responses. *Journal of Wood Science* 51(2): 136–140.
Shi, W. 2006. Research on the application mode and prospect of modern wood structure in China. *Nanjing Forestry University.*
Wang, P., Huang, K.Z. & Zhou, J.T. 1996. Relationship between the microcrack density and mechanical properties of marble. *Journal of Engineering Geology* 4(2): 19~23.
Wang, Z.H. & X.J. Yang. 2011. The overview of current status and development for China archetecture of madern wooden construction. *Woodworking Machinery* (02): 5–8.
Xu, S.L., Wu, W. & Li, T. 2001. Experimental studies on localization and bifurcation behaviors of a marble under triaxial compression. *Chinese Journal of Geotechnical Engineering* 23(3): 296~301.
Yin, Y.Q., Huang, J.F. & Wang, K.P. 1993. Experimental study of the constitutive behaviors of fragment marble. *Chinese Journal of Rock Mechanics and Engineering* 12(3): 240~248.
Yu, J. 2004. Wide prospects for wood structure building. *China Construction News.*
Zhou, X.Y., Cao, L., Zeng, D. & He, C.H. 2015. Flexural capacity analysis of glulam beams. *Building Structure* 45(22): 91-99.

Modern Engineered Bamboo Structures – Xiao, Li & Liu (eds)
© 2020 Taylor & Francis Group, London, ISBN 978-1-138-35185-1

Experimental research on glubam-concrete composite beams for bridges

T.Y. Li, B. Shan & Y.R. Guo
College of Civil Engineering, Hunan University, Changsha, China
Key Laboratory for Green & Advanced Civil Engineering Materials and Application Technology of Hunan Province, Changsha, China

ABSTRACT: This paper describes the outcomes of short-term tests on glubam-concrete composite (BCC) beams for bridge. Two types of connection have been tested: screw connector (SC) and notch connection (NC). Different variables include numbers and types of connection. Four-point bending tests to failure were performed on four full-scale BCC beams with 8.0 m long. All BCC beams with different connections exhibited satisfactory mechanical performance under short-term loading conditions in this investigation.

1 INTRODUCTION

Glubam, a new type of glue-laminated bamboo, was developed by Xiao et al. (2013). The short-and long-term experiments of glubam bridge show that glubam bridge can be used as a Truck load bridge and pedestrian bridge(xiao et al. 2010, xiao et al. 2014). For glubam-only flexural members, the main drawback of mechanical properties is relatively low stiffness, in order to meet the deformation requirements, the section height of the glubam beam will be large, which increases the self-weigh of glubam bridge, but also puts forward higher requirements for the processing of glubam beam. Therefore, improving the bending performance of glubam flexural members is a key issue for developing modern bamboo structures.

Based upon the existing technologies of the TCC system, developing glubam-concrete composite or bamboo-concrete composite (BCC) bridge or beam is a feasible solution for increasing strength and stiffness properties of glubam flexural members. In the system, an upper concrete slab is connected to a lower timber beam by shear connectors (Figure1), allowing the best properties of both materials to be exploited, combining the high compression resistance of concrete with the greater capacity of glubam to withstand bending and tension (Ceccotti 1995, Dias 2005, Lukaszewska 2009, Yeoh 2010, Fragiacomo et al. 2007). However, these advantages can be achieved only if the connection system provides structural composite action. A series of push-out tests had been conducted on total six types of connection and the results indicated that four of them were suitable for constructing the BCC systems, as previously reported by Shan et al. (2017). Therefore, the flexural performance of full-scale BCC bridges need to be investigated in order to present the design method, the full-scale bending tests to failure of BCC bridges, presented in this paper.

2 EXPERIMENTAL PROGRAM

2.1 *Specimens design*

Tow types of connection were selected for constructing the BCC bridges and conducting bending tests, as illustrated in Figure 1. (1) A screw with 18 mm diameter inserted 100 mm

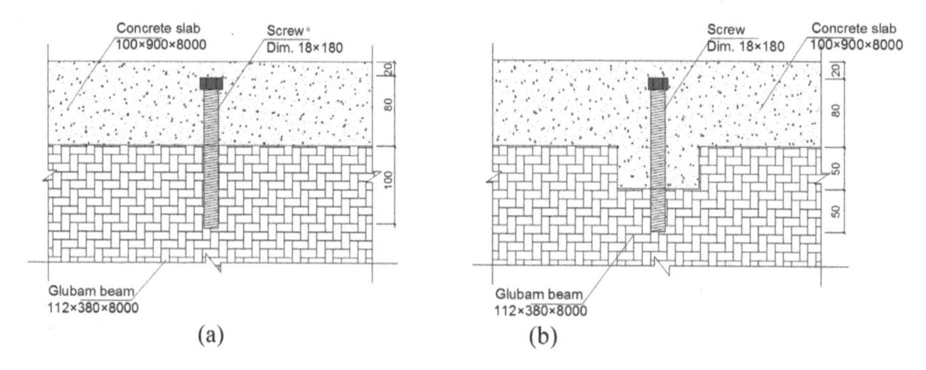

Figure 1. Details of connections (dimensions in mm): (a) Screw connector (SC); (b) notched connection (NC).

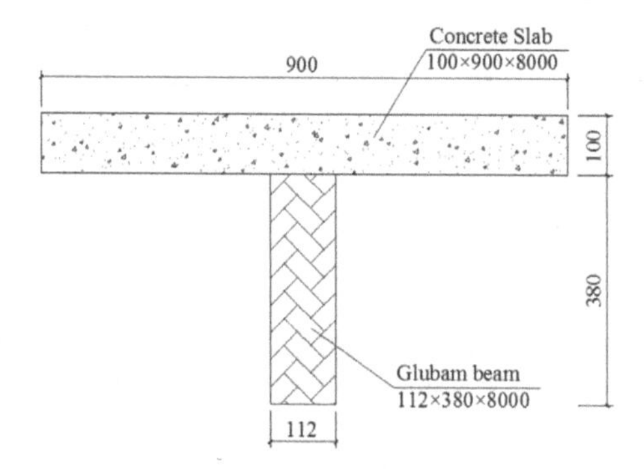

Figure 2. Cross-section diagram of BCC beams (dimensions in mm).

deep into the glubam (SC); (2) A rectangular notch reinforced with a screw of 18 mm diameter (NC).The details of each type of connection were presented in reference (Shan et al. 2017).

Four full-scale BCC beams with 'T' section were designed for 8000 mm length and have the same cross-sectional dimensions which consisted of a 900 mm × 100 mm (width × depth) concrete slab and a 112 mm × 380 mm (width × depth) glubam beam, as shown in Figure 2.

The details of each specimen are presented in Table 1 and Figure 3. The main experimental parameters include the type of connection, the number of connectors, and they can be classified as tow groups: (1) SC-25 and SC-45 used different number of screws as connector uneven arrangement along the longitudinal direction (Figure 3a ~ b.). (2) NC100-12 and NC100-16

Table 1. Details of all the BCC beams tested to failure.

Specimen	Type of connector	Number of connectors	Span (mm)	Height (mm)
SC-25	Screw connector	25	7800	480
SC-45		45	7800	480
NC100-12*	Notched connector (100 mm long)	12	7800	480
NC100-16		16	7800	480

*NC100-12 means that total 12 NC connectors with 100 mm long are used to connect concrete slab and glubam section in this specimen.

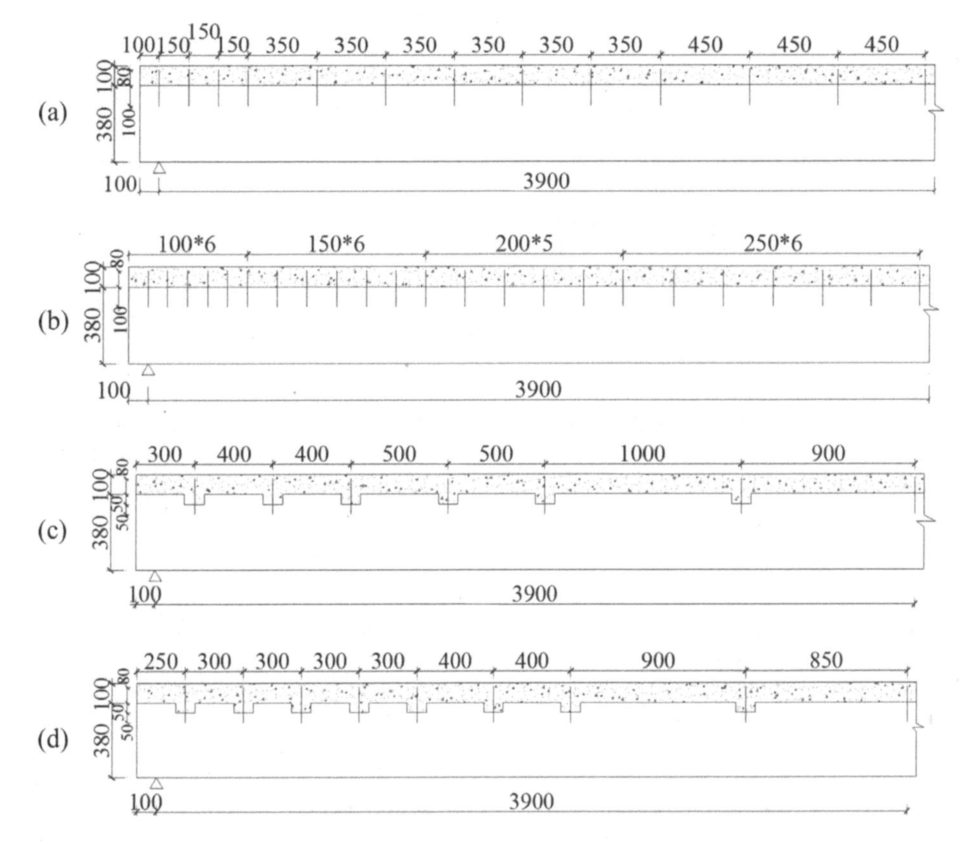

Figure 3. Details of the BCC beams (dimensions in mm): (a) SC-25; (b) SC-45; (c) NC100-12; (d) NC100-16.

were constructed with 100 mm long notches reinforced with a screw and arranged unevenly along the longitudinal direction (Figure 3c ~ d).

All connectors used in this research were glued in glubam beams by a two-component epoxy adhesive supplied by local manufacturer. These materials in this research, including glubam, concrete and adhesive, were the same to the materials used in the push-out test performed by Shan et al. (2017) and Wang (2018).

2.2 *Experimental set-up*

All beam specimens were simply supported and subjected to four-point bending tests to failure, according to the experimental setup displayed in Figure 4. The distance between

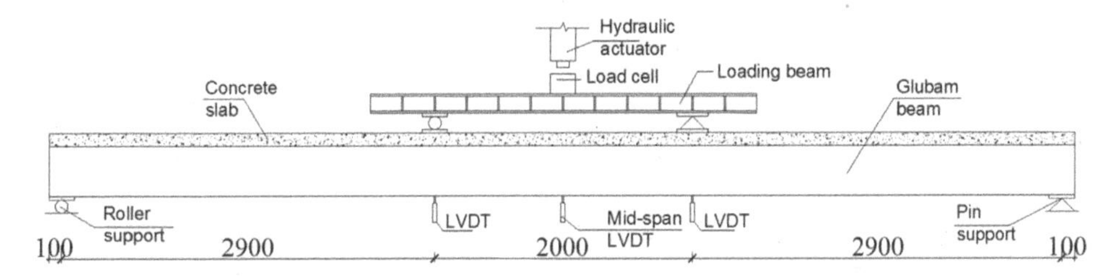

Figure 4. Details of test setup (dimensions in mm): Schematic diagram.

supports, L, was 7800 mm and the pure bending length was 2000 mm. The load was applied at the loading beam using a 400 kN hydraulic actuator and a spreader. The loading protocol followed during the test was similar to that recommended by EN 26891 (1991), with the bridge primarily loaded to 40% of the estimated failure load, held for 30 s, downloaded to 10% of the estimated failure load, held for 30 s and finally loaded up to the collapse of the beam.

During the tests, the following data were recorded: (1) the mid-span deflection of specimen. (2) relative slip between the concrete slab and the glubam beam at the first connector near the support of specimen. and (3) applied load.

3 TEST RESULTS AND ANALYSIS

The failure modes of all the beams are shown in Figure 5. Values of the main experimentally determined variables, for example, the failure load corresponding midspan deflections and end slips are summarized in Table 2. In addition, the total load versus midspan deflection are displayed in Figure 6.

3.1 *Failure modes*

Two SC specimens exhibited a brittle failure. Careful examination of the specimens after the tests revealed that the some bamboo strips were sheared and detached from glubam beam at the end (Figure 5a), and no plastic hinge formed in the screw at the glubam beam-concrete interface (Figure 5b). The results are possibly due to the poor quality of glubam sheets in the factory.

The failure hierarchy observed for specimens of NC series is fracture in tension of glubam at the support region under the loading beam, together with crushing of concrete with plasticization of the screws in the notched connections observed (Figure 5c ~ d).

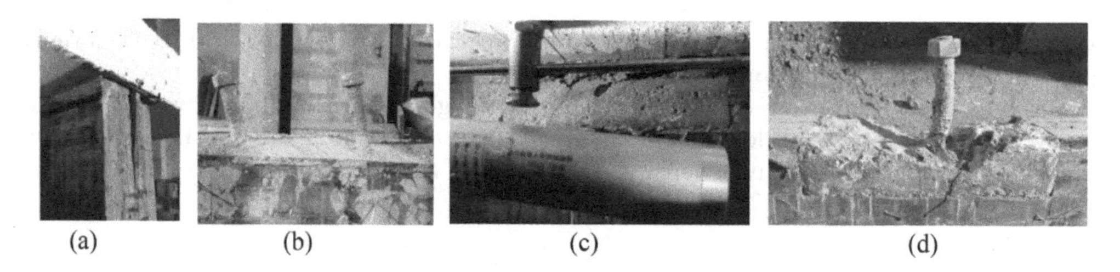

(a) (b) (c) (d)

Figure 5. Failure modes: (a) Shearing damage at the end (SC-25); (b) deformation of screws (SC-25); (c) concrete cracking in notch (NC100-12); (d) plastic deformation of screw (NC100-12).

Table 2. Experimental results of the tests to collapse performed on BCC beams.

Specimen	$2P_{max}$* (kN)	Δ_{max} (mm)	s_{max} (mm)
SC-25	128.6	114.6	4.113
SC-45	166.4	109.9	2.776
NC100-12	107.6	95.7	5.842
NC100-16	123.3	84.7	3.902

*$2P_{max}$ corresponds to the maximum resultant of two point loads.

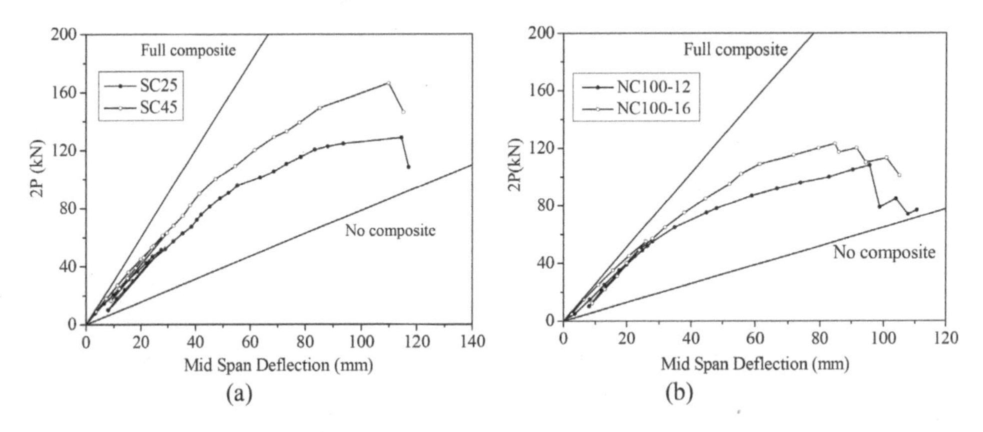

Figure 6. Load-midspan deflection responses of specimens: (a) SC series; (b) NC100 series.

3.2 *Load-midspan deflection curves*

Figure 6 illustrates relationships of the experimental total load, corresponding to the resultant of two point loads, $2P$, versus midspan deflection. The experimental curves are shown within two limit curves representing the estimated behavior of the glubam-concrete system with perfect rigid connections (the upper limit) and no connections (the lower limit) between the concrete slab and glubam beam. It can be seen that compared with glubam bridges (xiao et al. 2010, xiao et al. 2014), the stiffness and bearing capacity of glubam-composite beams for bridge are greatly increased, which can significantly reduce the section height of glubam beams, or obtain a greater bearing capacity with the same section height of glubam beams.

Figure 6a displays the load-deflection response of SC series beams, where the collapse loads of SC-25 and SC-45 were 128.6 kN and 166.4 kN, corresponding the midspan deflections of 114.6 mm and 109.9 mm, respectively. From this diagram, the load increased slowly up to the peak load followed by a sudden reduction, due to the poor quality of glubam sheets. The composite beams exhibit limited plastic deflection, despite the excellent ductility of this type of connector in previous shear tests (Shan et al. 2017).

Four NC composite beams present a similar load-deflection relationship, which behaved linear feature at the initial stage, and then slight nonlinearity was detected beyond the serviceability limit state (SLS) corresponding to the deflection reaching *L*/300. Up to the failure of the system, there was a considerable recovery of strength after the peak load, especially for the specimens of NC100-16, as shown in Figure 6b.

3.3 *Horizontal slip between concrete slab and glubam beam*

The relative slip between concrete and glubam was monitored at the connector locations using LVDTs. The test data show that the largest slip normally occurred in the connector nearest to the end or support, and the value of slip decreases gradually from the end to midspan region (Wang 2017). The maximum relative slip between concrete and glubam at collapse is listed in Table 2.

Comparisons of the slip values at collapse of the first connector in each specimen are also displayed in Figure 7, which shows the load-slip curves obtained from the push-out tests (Shan et al. 2017). Generally, a ductile failure pattern is defined as the case where a connector can withstand a relative slip of 10 mm with the shear resistance decreased not more than 20% of the peak value (Deam et al. 2007). So the ultimate point of connector (s_u) is defined as the point corresponding to the relative slip reaching 10 mm for the ductile failure or 80% of the maximum load during the decline stage of the load-slip curve for the brittle failure, respectively (Dias et al. 2011).

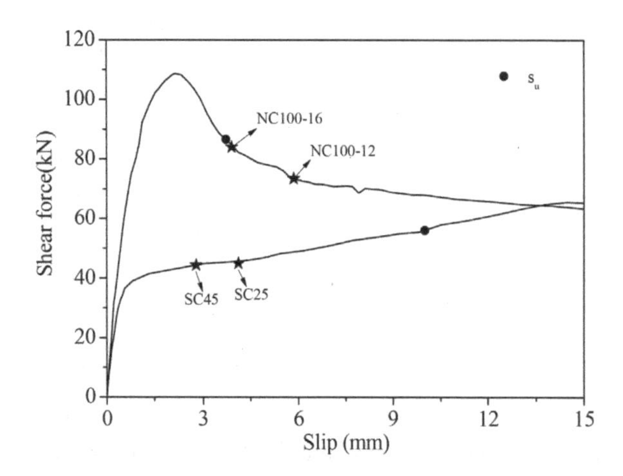

Figure 7. Experimental load-slip curves of each connections.

Figure 7 presents the load-slip curves of each connection gained from push-out tests (Shan et al. 2017, Wang 2018), and the ultimate slips at the first connector near the support of specimen are also marked. It is observed that the ulitmate slips of SC series are significantly smaller than 10 mm, providing that the connectors were hardly plasticized at collapse of the beams. Therefore, ductile behavior of the entire composite beams was prevented by the brittle failure of glubam due to the poor quality of some of glubam sheets. Contrary to the SC series, the ultimate slips of NC series at collapse are obviously larger than the ultimate point (s_u), meaning that the BCC specimens can carry more load even after the failure of the first connector. It is due to the redistribution of shear force at the interface caused by a consecutive failure from the first connector to others. This result indicates that the BCC beam connected by brittle connectors also can exhibit good deformation capacity, corresponding to the analysis result reported by Lukaszewska et al. (2010).

4 CONCLUSIONS

All BCC beams for bridge with different connections exhibited satisfactory mechanical performance under short-term loading conditions in this investigation. On the basis of the results obtained from tests and analysis, the following general conclusions can be drawn.

For SC series, over ductility of composite beams was prevented by the use of glubam sheets with the poor quality before the ductile connection could plasticize. Possible ways to increase the ductility of the system include the use of stronger glubam material, or larger glubam cross sections to delay the failure of the glubam beam.

For NC series, the use of more connectors (or decreasing the spacing of connectors) can make sure of a considerable strength recovery after peak load. This is an important outcome as it ensures a moderate ductile behavior of the composite beam which may allow sufficient time for evacuation in the case of an emergency.

In general, compared with glubam bridges, glubam-concrete composite beams can be used for medium and small-span Truck load bridges and pedestrian bridges.

ACKNOWLEDGEMENTS

The experimental work of this research was conducted at the Center for Integrated Protection Research of Engineering Structures, the MOE Key Laboratory of Building Safety and Efficiency at Hunan University under the supports of the National Key R&D Program of China (2017YFC0703502).

REFERENCES

B, Shan. & Y, Xiao. & W.L, Zhang. & B, Liu. 2017. Mechanical behavior of connections for glubam-concrete composite beams. Construct Build Mater 143(2017): 158–168.

Ceccotti, A. 1995. Timber-concrete composite structures. in Timber Engineering, Step 2,1st Ed. Centrum Hout, The Netherlands, E13/1-E13/12, p. 16.

Committee European Normalisation. 1991. Timber Structures Joints Made with Mechanical Fasteners-General Principles for the Determination of Strength and Deformation Characteristics. EN 26891, Brussels, Belgium.

Deam, B.L. & Fragiacomo, M. & Buchanan, A. 2007. Connections for composite concrete slab and LVL flooring systems. Mater Struct J. ISSN 1871-6873.

Dias, A.M.P.G. 2005. Mechanical behavior of timber-concrete joints. in Technische Universiteit, Delft, The Netherlands.

Dias, A.M.P.G. & Jorge, L.F.C. 2011.The effect of ductile connectors on the behavior of timber-concrete composite beams. Eng Struct 33: 3033–3042.

Fragiacomo, M. & Amadio, C. & Macorini, L. 2007. Short-and long-term performance of the "Tecnaria" stud connector for timber-concrete composite beams. Mater Struct 40 (10): 1013–1026.

Lukaszewska, E. 2009. Development of prefabricated timber-concrete composite floors.Lulea (Sweden): Department of Civil, Mining and Environmental Engineering, Division of Structural Engineering, Lulea University of Technology.

Lukaszewska, E. & Fragiacomo, M. & Johnsson, H. 2010. Laboratory tests and numerical analysis of prefabricated timber-concrete composite floors. J Struct Eng 136(1): 46–55.

Y, Xiao. & Q, Zhou. & B, Shan. 2010. Design and construction of modern bamboo bridges. J. Bridge Eng 15(5): 533–541.

Y, Xiao. & R.Z, Yang. & B, Shan. 2013. Production, environmental impact and mechanical properties of glubam. Construct Build Mater 44(2013): 765–773.

Y, Xiao. & L, Li. & R.Z, Yang. 2014. Long-term loading behavior of a full-scale glubam bridge model. J. Bridge Eng 19 (9):04014027.

Yeoh, D. 2010. Behaviour and design of timber-concrete composite floor system. Christchurch (New Zealand): Department of Civil and Natural Resources Engineering, University of Canterbury.

Z.Y, Wang. 2018. Experimental tests and analysis of Glubam-concrete composite beams (Dissertation), Hunan University, Changsha.

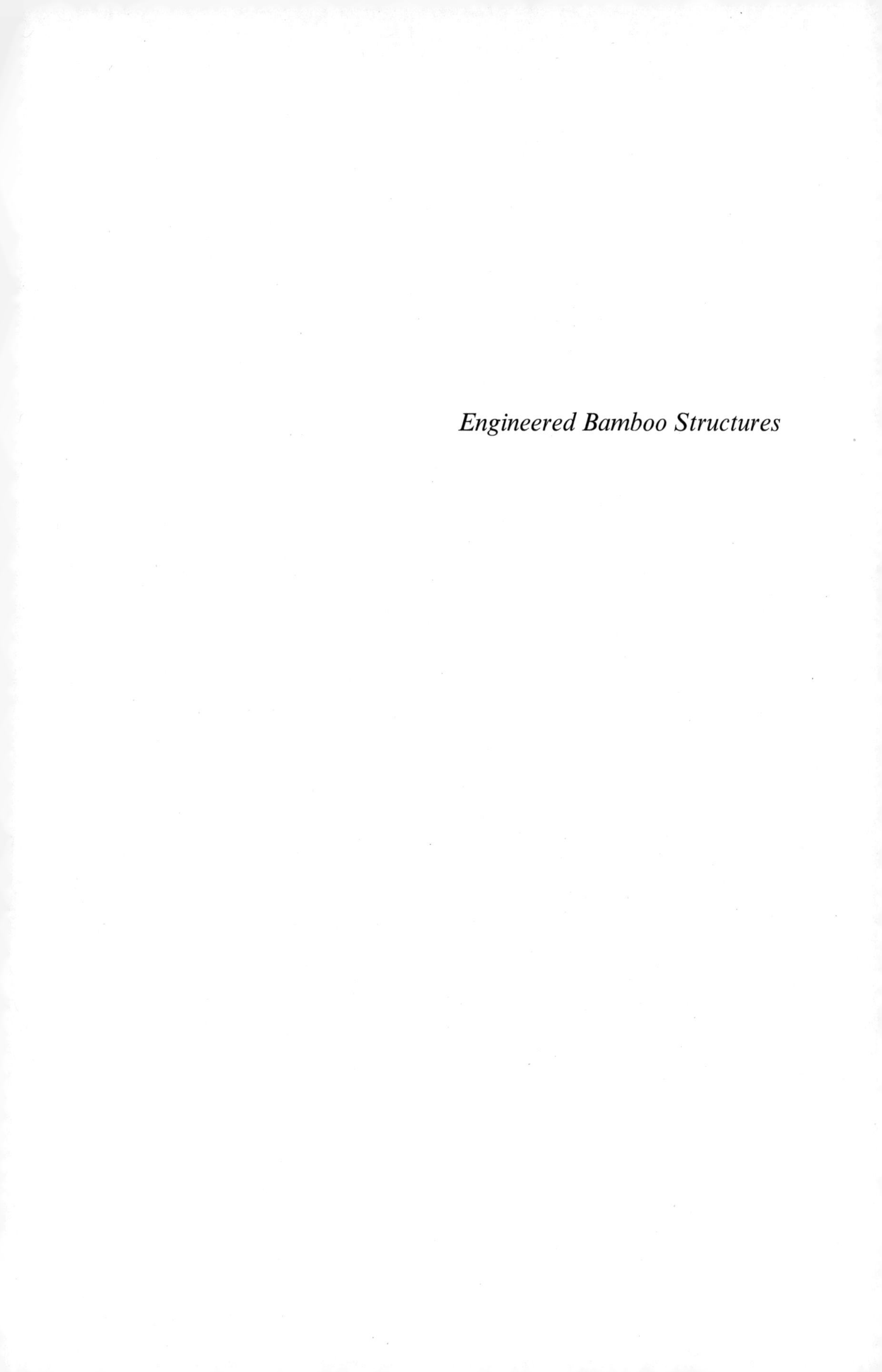

Engineered Bamboo Structures

Modern Engineered Bamboo Structures – Xiao, Li & Liu (eds)
© 2020 Taylor & Francis Group, London, ISBN 978-1-138-35185-1

Lightweight glubam structures manufactured using industrialized ply-bamboo panels

R. Wang
Hunan University, Changsha, Hunan, China

Z. Li
Nanjing Tech University, Nanjing, Jiangsu, China

G. Chen
College of Civil Engineering, Nanjing Forestry University, Nanjing, Jiangsu, China

Y. Xiao
Zhejiang University-University of Illinois Institute (ZJUI), Zhejiang University, Haining, Zhejiang, China

ABSTRACT: One or two stories lightweight glubam structures based on industrialized ply-bamboo panels, along with corresponding mechanical researches have been introduced in this paper. The main structural elements, especially the lightweight frame shear wall with its metal connections, and its mechanical performance have been studied in recent 10 years. The modelling and design of such structure can be based on connection system information. The performance of a prototype house from 2009 to 2019 were also reported at the end of this paper. Existing researches indicated that the lightweight glubam structures has large potential to be widely used as residential or small commercial structures in city or rural regions.

1 INTRODUCTION

Lightweight wood frame constructions are used widely in residential and industrial buildings in North America for their good ability to resist seismic and wind loads, as well as good architectural performance. In China due to the limitations of timber resources and forest protection law, very small proration of new constructions are timber structures when compare with concrete and steel structures. Therefore, to find another type of environmental-friendly construction material which are abundant in China, meanwhile have good structural performance similar to wood is the main reason to stimulate the research performed herein. Bamboo is a possible solution. Since 1980s, following the advancement of bamboo industry, glued laminated bamboo products appeared in construction industry. A new type of bamboo materials, with a trademark of Glubam, has been studied by the last authors' research team since 2005, with a relatively systematic database established (Xiao & Shan, 2013, Xiao et al., 2014).

2 MATERIALS

Gluabm is structural used glued laminated bamboo made by 30-40 mm thickness industrialized bamboo panel products. These types of bamboo panel products can be further cold-glued to form various dimensions of glubam elements and to meet various structural and architectural requirements. There are two types of industrialized ply-bamboo panel products widely used in lightweight glubam constructions up to now, which are thick strip ply-bamboo panels laminated

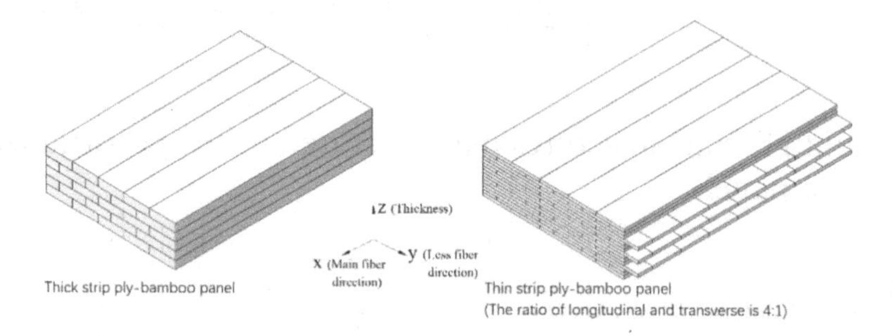

Figure 1. Configuration of ply-bamboo panels for glubam.

Table 1. Mechanical properties of ply-bamboo panels.

Property			Mean (MPa)	SD (MPa)	COV (%)
Thick strip ply-bamboo panel (Li et al., 2019)					
Tension	Main fiber direction		85	12.8	16.9
Compression	Main fiber direction		73	3.4	4.6
	Less fiber direction		24.8	0.9	3.5
Bending	MOE	Bending around y axis	10500	1021.1	9.3
	MOR		108.9	9.2	8.4
	MOE	Bending around z axis	11200	757.7	6.4
	MOR		119.5	8.7	7.2
Shear	Parallel to glue line τ_{xz}		16.9	2.9	9.3
Thin strip ply-bamboo panel (Xiao et al., 2013)					
Tension	Main fiber direction		83	16	20
Compression	Main fiber direction		51	2.6	5.0
	Less fiber direction		25	2.9	12
Bending	MOE	Bending around y axis	4900	598	27
	MOR		24	6	25
	MOE	Bending around z axis	9400	943	10
	MOR		99	12	12
Shear	Parallel to glue line τ_{xz}		16	2	12

Note: COV: coefficient of variance; MOE: modulus of elasticity; MOR: modulus of rupture; SD: s coordinate system is shown in Figure 1.

by bamboo strips of about 5-8 mm thick and thin strip ply-bamboo panels laminated by bamboo strips of about 2 mm thick, as shown in Figure 1. The bamboo fibers in thick strip ply-bamboo panels are normally all along longitudinal direction, while thin strip ply-bamboo panels are bidirectional, the ratio of longitudinal and transverse strips is variable. The main mechanical properties of these two types of panels are shown in Table 1.

3 MAIN COMPONENTS

Similar to lightweight wood-frame construction, the lightweight glubam structure can be constructed following the so-called platform construction, as shown in Figure 2. The floor of each story is constructed of joists covered with sub-flooring to form a working surface upon which shear walls are erected and materials are stacked. The shear walls are connected to the foundation or the walls in lower story through the floor system by anchor connections. The roof system, normally prefabricated trusses, is connected to the top beam of shear walls by

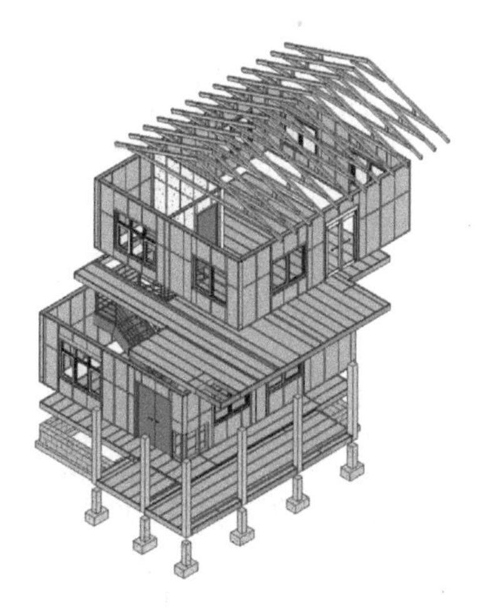

Figure 2. Lightweight glubam building under construction (Chen, 2011) and BIM model.

metal connections. The lightweight glubam frame structures have good lateral resistant behavior. The shear walls perpendicular to the lateral force can transfer the load to the horizontal diaphragms (floor or roof), then the horizontal diaphragms distribute it to the shear walls parallel to the lateral force, finally the shear walls transfer the force to the foundation.

3.1 *Shear wall and its connection system*

As lightweight wood-frame shear walls, the lightweight glubam shear wall is the main component to resist lateral loads. Figure 3 describes the fabrication process and its structural elements and connection system. The connection system consists of panel-frame connection, frame-frame connection and hold-down connection, which is a dominant factor to determine the lateral performance of glubam shear wall. The research of the connection system was

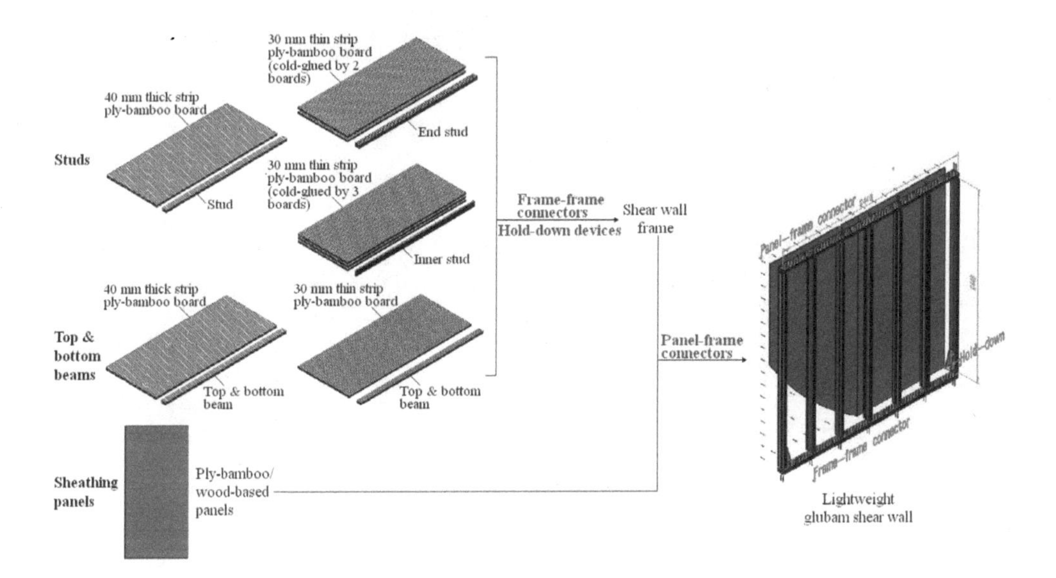

Figure 3. Fabrication process of glubam shear wall and connection system.

conducted by Wang et al., 2019, and based on the test information of connection system, finite element models can be established to simulate the lateral performance of glubam shear walls. Existing studies (Wang et al., 2017, Wang et al., 2019) indicates that lightweight glubam shear walls have high bearing capacity but limited ductile performance than wood-frame shear walls when using the same panel-frame connectors.

3.2 *Floor diaphragm and roof system*

The floor system of lightweight glubam structure consists sills, girders, joists or floor trusses and sub-flooring. The joists usually is spaced 305 mm, 405 mm or 610 mm on centers and supported by the foundation and the center girder. The size of girders is determined by load calculation and detailing requirements. Sub-flooring, or flooring sheathing is applied over the framing to provide a working platform, and common sheathing material is 15 mm thin strip ply-bamboo panel.

The roof system is typically made with prefabricated glubam trusses. As engineered trusses can reduce on site labor and provide greater flexibility in the layout of interior walls. Two types of glubam roof trusses were tested by Xiao et al., 2014. The experimental results showed that glubam trusses had adequate stiffness and strength, the failure mode was lateral buckling of the top chords. Xie and Xiao, 2016 did some research on large-span glubam roof trusses. Two full-scale 20 m long trusses were studied, one was typical glubam truss and the other was hybrid truss with glubam elements as top chords and high strength steel bars as lower chords. For large-span truss, the results indicated

Figure 4. Floor diaphragm in lightweight glubam structures (Chen, 2011).

Large-span glubam truss | Large-span glubam-steel truss | Glubam-steel space truss

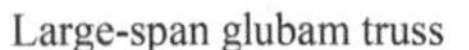

Figure 5. Roof & canopy trusses.

that the hybrid truss was superior to the glubam one. Glubam and steel hybrid space truss also researched by Wu & Xiao, 2018.

4 TEN-YEARS PERFORMANCE OF A PROTOTYPE HOUSE

The first lightweight glubam frame prototype house was built in Feb. 2009 and is in the College of Civil Engineering of Hunan University with 250 m^2 of building area. This house has experienced extreme weather and natural disaster for many times and experienced a horizontal removal in 2013. All of the structural elements were prefabricated in the factory and transported to the construction site. To study the seismic behavior of this residential

(a) Lightweight glubam frame house built in 2009

(b) Relocation of world glubam house in 2013

(c) Lightweight glubam frame house in 2019

Figure 6. Lightweight glubam frame house in Hunan University.

house, shake table tests were conducted using one room unit of first floor, with the size of 2.44 m×3.66 m in plane and 2.6 m in height (Chen et al., 2011). Test results showed that the room model had no visual damage for 0.3 g PGA (peak ground acceleration) because of lower weight. For 0.4 g PGA, only the edge and opening of shear walls had minor damage. The room model could withstand under action of three different seismic ground motions within a range of 0.5 g PGA, and the observed damage remained noncritical. The room model used in shake table test was reconstructed with finishing of the inside surfaces to study the fire safety of the house (Xiao & Ma, 2012). After one hour of exposure to fire inside the model, the result showed good fire resistance of glubam elements. The walls and the ceilings were essentially intact, the existence of carburization layer under fire can delay further penetrating of fire into the structure.

5 CONCLUSIONS

Lightweight glubam structures based on industrialized bamboo panel products were introduced in this paper. Main structural elements consist of lightweight glubam shear walls, floor diaphragm and roof system. The research of material properties and structural elements were conducted in recent 10 years. A key determining factor governing the performance of such structure is connection system. Design and simulation of this structure can be based on the information of connection system. The research and the prototype house indicate that glubam material has good mechanical properties, low impact on environment and lightweight glubam structures have better structural performance and durability.

REFERENCES

Chen, G. 2011. Experimental study and engineering application of light frame bamboo structure. Ph. D. thesis, Hunan University: Changsha, China. (in Chinese)

Chen, G., Shan, B., Xiao, Y. 2011. Aseismic performance tests for a light glubam house. *Journal of Vibration and Shock* 30(10):136–142. (in Chinese)

Li, Z., Yang, G.S., Zhou, Q., Shan, B., Xiao, Y. 2019. Bending performance of glubam beams made with different process. *Advances in Structural Engineering* 22(2):535–546.

Wang, R., Xiao, Y., Li, Z. 2017. Lateral loading performance of lightweight glubam shear walls. *Journal of Structural Engineering* 143(6).

Wang, R., Wei, S.Q., Li, Z., Xiao, Y. 2019. Performance of connection system used in lightweight glubam shear wall. Construction and Building Materials 206:419–431.

Wu, Y. & Xiao, Y. 2018. Steel and glubam hybrid space truss. *Engineering Structures* 171:140–153.

Xiao, Y. & Ma, J. 2012. Fire simulation test and analysis of laminated bamboo frame building. *Construction and Building Materials* 34:257–266.

Xiao, Y. & Shan, B. 2013. Glubam structures. China Architecture & Building Press: Beijing, China. (in Chinese)

Xiao, Y., Yang, R.Z., Shan, B. 2013. Production, environmental impact and mechanical properties of glubam. *Construction and Building Materials* 44:765–773.

Xiao, Y., Chen, G., Feng, L. 2014. Experimental studies on roof trusses made of glubam. *Materials and structures* 47:1879–1890.

Modern Engineered Bamboo Structures – Xiao, Li & Liu (eds)
© 2020 Taylor & Francis Group, London, ISBN 978-1-138-35185-1

Image-based 3d reconstruction of a glubam-steel spatial truss structure using mini-UAV

G. Candela & V. Barrile
Mediterranea University of Reggio Calabria, Reggio Calabria, Italy

C. Demartino & G. Monti
Nanjing Tech University, Nanjing, Jiangsu, China

ABSTRACT: During the last years, thanks to the availability of commercial and low-cost UAVs (Unmanned Aerial Vehicles), surveying by means of 3d reconstruction generated by computer vision algorithms is having a rapid growth. These tools allow for a quick and effective methodology to acquire image data and reconstruct a detailed 3d model that can be used for structural analysis and maintenance operations. In this context, this paper presents an application example of a commercial mini-UAV equipped with a low-cost camera in an aerial photogrammetric survey of a bamboo-steel spatial truss roof system, with the purpose of obtaining the as-built drawings. This roof system is located at the entrance of the College of Civil Engineering of the Nanjing Tech University and is made by a spatial truss system supporting a glass roof in which the GluBam beams are used for the upper chord and the diagonal elements, while steel tubes are used for the lower chord. The developed methodology for acquiring data through UAV is introduced, specifying the flight acquisition plan to obtain the images data, and the data processing using structure-from-motion algorithms to provide an accurate 3d-model. The accuracy of the measurements is evaluated by comparing results with project drawings. Moreover, the use of the obtained 3d model and photo dataset in practical application is discussed. The results highlight the economical and fast applicability of this technology to the survey of spatial truss structures to derive a 3d model, which can also be used to acquire useful information on the structural health conditions.Surveying techniques for the acquisition of spatial information are rapidly growing during the last years thanks to the availability of new technologies and instruments (both software and hardware). The digitalization of the construction industry is revolutionizing the design process and the entire life cycle, in particular the maintenance process. Modern surveying techniques are fundamental to acquire spatial information and convert it into digital information, in case blueprints are not available, and also to analyse and monitor the structure and to schedule maintenance works.

1 INTRODUCTION

In this context, the aerial survey, applying photogrammetric techniques, of a spatial truss made of bamboo-steel is presented. This material is used as structural material in Asia, South Africa and South America thanks to its large availability and sustainability. The survey was realized using images taken from an Unmanned Aerial Vehicle (UAV), so that the entire 3d scene, with the measurable model of the roof system, was reconstructed.

In the first part of this paper, the structure is presented with technical details; subsequently, modern techniques for surveying and reconstruct 3d model starting from photos and using aerial photogrammetry are described. Finally, the obtained 3d model is compared with the blueprints for the results validation. Other possible application and improvements are described in the conclusions.

2 GLUBAM – STELL SPATIAL TRUSS STRUCTURE

2.1 *Bamboo as a structural material*

The analysed structure is the roof system located at the entrance of the Civil Engineering College on Nanjing Tech University in Nanjing, China (Figure 1). The applications of bamboo in civil engineer as structural material are several (Sharma *et al.*, 2015), thanks to its diffusion and sustainability (Escamilla *et al.*, 2014). Its most appealing features are attributable to the fact that bamboo is a highly renewable construction material with low embodied energy and high strength-to-weight ratio.

The use of structural engineered bamboo composites to replace timber in buildings is rapidly growing in Asia, also thanks to the good equivalent thermal performance (Wang *et al.*, 2018). The analysed truss system structure is made of glubam, a composite bamboo material whose characteristics and production process are described in (Xiao *et al.*, 2013).

2.2 *Surveyed spatial truss description*

The structure of the roof system, built in 2017, is a spatial composite truss made of glubam and steel that measure 12000 mm × 3600 mm (Figure 2), and it is composed by 10 × 3 identical square pyramids, with vertex on the bottom chord, having base of 1200 mm × 1200 mm and height of 849 mm (Figures 3, 4). The in-plane dimension of the bottom part of this spatial structure is 10800 mm × 2400 mm.

The structure is simply supported, with the three nodes on the second alignment constrained by hinges and the three nodes on the last but one alignment resting on rollers.

The upper chord and the diagonal bars are made of glubam, whereas the lower chord is made of steel tubes. The bars made of glubam have square cross-section of 56 mm × 56 mm (Figure 5).

These bars are manufactured by glueing 9 smaller square elements (each one composed by 3 or 4 thin bamboo strips) through a 3 × 3 arrangement. The steel tubes of the lower chord have an external diameter and wall thickness equal to 42 mm and 4 mm, respectively.

Figure 1. Bamboo steel spatial truss structure in Nanjing Tech University (Wu and Xiao 2018).

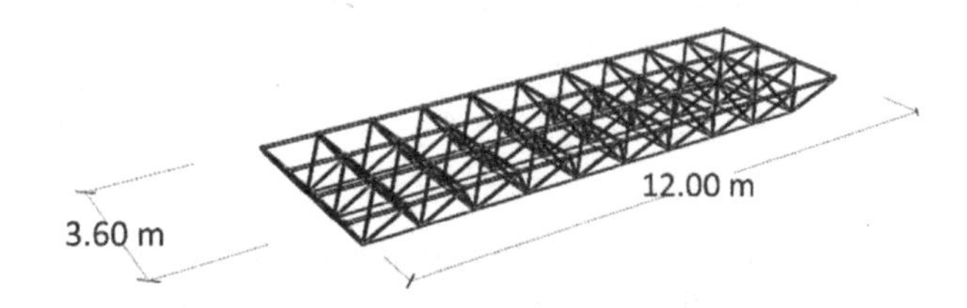

Figure 2. 3d view of the spatial truss structure.

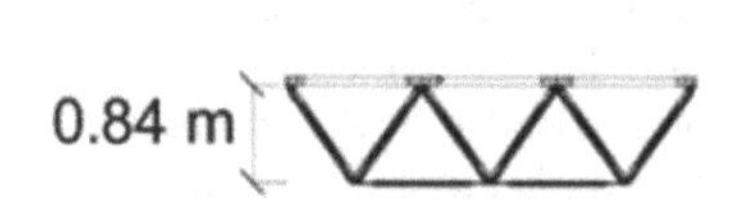

Figure 3. Side view of the spatial truss.

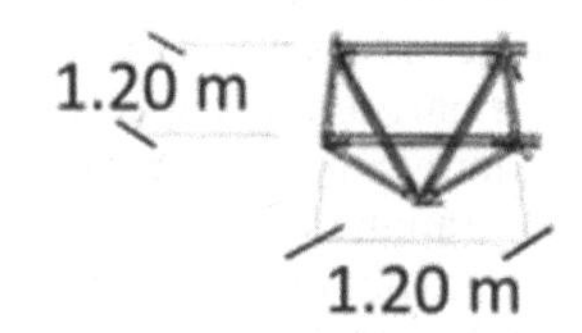

Figure 4. Pyramids module.

Figure 5. Glubam bars and cross section.

3 IMAGE-BASED 3D RECONSTRUCTION USING UAV

3.1 *Close range and aerial photogrammetry*

The application of photogrammetry for surveying and 3d reconstruction is possible thanks to the availability of low-cost sensors and the continuous improvement of computer vision algorithms to elaborate the images. Nowadays these algorithms, based on Structure From Motion (SFM) (Faugeras *et al.*, 2001) (Wu, 2013), have reached a good level of maturation and are available in different commercial software. The applications of this techniques are various (Westoby *et al.*, 2012), from civil engineering (Leonardi *et al.*, 2018) to cultural heritage preservation (Yilmaz *et al.*, 2007). In particular, close-range photogrammetry (Luhmann *et al.*, 2013), a photogrammetric technique where the camera is placed on the ground, is used to acquire metrics information of small objects, and also to reconstruct three-dimensional measurable scenes (Miller, 2009). In aerial photogrammetry, instead (Colomina *et al.*, 2014), the camera sensor is mounted on a UAV system, that makes the process really fast and replicable (Barrile *et al.*, 2017) and allows also to reach inaccessible areas with different angles.

Consequentially, compared with other traditional techniques, aerial photogrammetry survey is used to quickly acquire spatial and metric information thanks to the possibility of recreating the measurable 3d scene affordably and quickly.

3.2 *3d reconstruction from images workflow*

The correct acquisition of data is the first step in the workflow for 3d reconstruction. The quality of the reconstruction highly depends on the quantity and quality of the acquired photos. However, increasing the number of photos implies an increase of the calculation time. The minimum overlap between two consecutive photos to obtain a good reconstruction is set on 70% on both vertical and horizontal side.

The creation of the 3d model from image dataset is based on SFM algorithms. Starting from photos, the Scale Invariant Feature Transform (SIFT) (Lowe, 1999) recognizes the unique feature on each photo, while the SFM algorithm matches common features on each photo, calculates the camera positions, and recreates the target scene or object as a 3d point cloud. Subsequently, the point cloud is cleaned and simplified, reducing the number of points through Poisson Disk Algorithm (Corsini *et al.*, 2012) and the triangular mesh is reconstructed using Screened Poisson Surface Reconstruction Algorithm (Kazhdan *et al.*, 2013).

The output of this process is a 3d triangular mesh of the target scene with a texture on the objects.

4 CASE STUDY: SURVEYING AND MODELLING OF A GLUBAM STRUCTURE

4.1 *Data acquisition plan*

The survey of the glubam-steel spatial truss was performed using aerial photogrammetry with a commercial UAV DJI Mavic Pro (DJI, Shenzhen, China).

This UAV uses a high-resolution camera with built-in GPS in order to geotag each photo, saving the positions information in an EXIF file. Its low weight (700 gr) allows to perform accurate survey ensuring centimetric precision at a low distance (5-10 m) from the object; moreover, it represents a fast option to acquiring spatial information of the target scene.

The characteristics of the camera sensor mounted on the UAV are reported in Table 1.

The flight plan was set to obtain a Ground Sampling Distance (GSD, i.e., the average distance in the photo between the centre of two consecutive pixels) of 0.5 cm/pixel on each photo and 80% overlap between two consecutive photos.

The UAV acquisition plan was set using the commercial software Pix4d (Pix4d SA, Lausanne, Switzerland) in free-flight mode, acquiring a photo on a grid of 1 m × 1 m both vertically and horizontally. To obtain the required GSD, three different flights in correspondence of the joints and on the glubam pieces were performed. A total of 459 photos (2.5 GB of files) were acquired in 30 minutes of survey.

4.2 *3d reconstruction process*

The obtained dataset was then elaborated with the commercial software 3d Flow Zephyr (3Dflow srl, Verona, Italy) that integrates the SFM algorithms and Poisson Surface Reconstruction Algorithm to generate, respectively, the 3d point cloud and the 3d mesh (Figure 7) of the target scene.

To ensure the metric precision of the reconstruction, the entire model was scaled with a real measure of the object. In Table 2 the data obtained from the 3d reconstruction process are shown.

4.3 *Measurements and comparison with BIM model*

The acquired model was then simplified using the 3d computer graphics software Rhinoceros 6 (Robert McNeel & Associates, Seattle, Washington, USA) together with the Mesh 2 Surface plug-in (KVS OOD, Blagoevgrad, Bulgaria) providing simple and efficient tools to convert

Table 1. DJI Mavic Pro camera characteristics.

Sensor	Sony FC220 – 1/2/3" CMOS
Image resolution	12,35 megapixel
Lens	28 mm f/2.2
Real focal length	5 mm
Real sensor width	6.7 mm
ISO range	100-1600
Electronic shutter speed	8 s – 1/8000 s
Image size	4000 × 3000 pixel
Geotagging	Built-in GPS

Figure 6. Camera position in the target scene.

Figure 7. 3d mesh of the roof system.

Table 2. Reconstruction process data workflow.

N. of photos	459 (2.5 gb)
3d point cloud	3 million points
Running time	5 hours
3d mesh	839.405 points - 1.674.921 triangles
Running time	1 hour

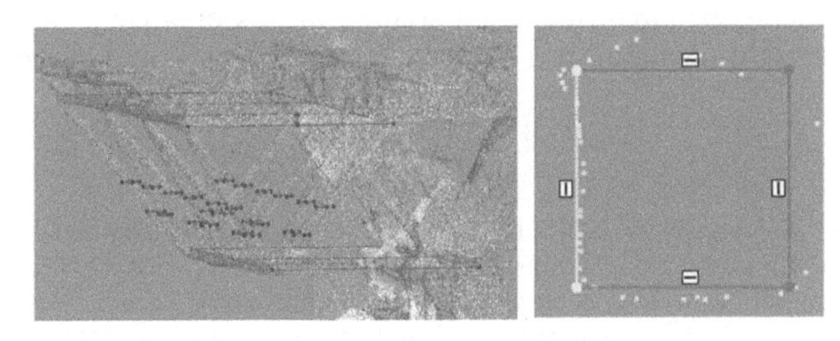

Figure 8. Cross section of the 3d mesh.

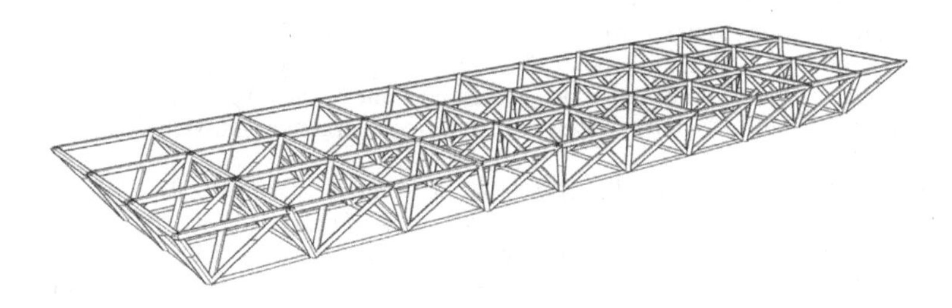

Figure 9. 3d model obtained from UAV survey.

a 3D scanned mesh into a CAD model. Three different sections of the glubam bars were performed (Figure 8), approximating the cross-section to obtain metric information and measurement and a simplified model based on the original mesh.

Based on the extracted section, a comparison between the blueprints and the obtained model was carried out, assessing a measure difference of 2 mm (58 mm of the reconstructed model and 56 mm in the real one).

The reconstructed model is represented in Figure 9 and the complete measurable 3d model of the glubam-steel truss is shown in Figure 10.

Figure 10. Photos and 3d model of roof system.

5 CONCLUSIONS

The acquisition of a roof system in bamboo-steel using a UAV has been presented. This acquisition was performed using aerial photogrammetry to obtain a complete measurable 3d model of the glubam-steel truss, with a fast and affordable procedure. The obtained results were compared with blueprints to validate the 3d mesh model. An additional feature of this methodology, which is being explored in current studies, allows obtaining, from image analysis, useful information on the health of the structure, of bamboo parts and of joints.

REFERENCES

Barrile, V., Bilotta, G. & Nunnari. A. 2017. 3D modeling with photogrammetry by UAVs and model quality verification. In: 4th international workshop on geoinformation science: GeoAdvances 2017. ISPRS annals of the photogrammetry, remote sensing and spatial information sciences, Vol. 4, p. 129–134.

Colomina, I., Molina, P. 2014. Unmanned aerial systems for photogrammetry and remote sensing: A review. *ISPRS Journal of Photogrammetry and Remote Sensing*.

Corsini, M., Cignoni, P. & Scopigno, R. 2012. Efficient and flexible sampling with blue noise poperties of triangular meshes. *IEEE Transaction on Visualization and Computer Graphics* 18 (6), 914–924.

Escamilla, E.Z., Habert, G. 2014. Environmental impacts of bamboo-based construction materials representing global production diversity. *Journal of Cleaner Production* 69: 117–127.

Faugeras, O.D., Luong, Q.T. & Papadopoulo, T. 2001. The geometry of multiple images - the laws that govern the formation of multiple images of a scene and some of their applications. DBLP.

Kazhdan, M., Hoppe, H. 2013. Screened poisson surface reconstruction. ACM Transactions on Graphics (TOG), 32 (3), 29.

Leonardi, G., Barrile, V., Palamara, R., Suraci, F. & Candela, G. 2018. Road degradation survey through images by drone. *Smart Innovation, Systems and Technologies* (Vol. 101).

Lowe, D.G. 1999. Object recognition from local scale-invariant features. IEEE.

Luhmann, T., Robson, S., Kyle, S. & Boehm. 2013. Close range photogrammetry and 3D imaging.

Miller, P.E. 2009. Applications of 3D measurement from images. *The Photogrammetric Record.*

Sharma, B., Gatóo. A., Bock, M. & Ramage, M. 2015. Engineered bamboo for structural applications. *Construction and Building Materials* 81: 66–673.

Wang, J.S., Demartino. C., Xiao, Y. & Lib, Y.Y. 2018. Thermal insulation performance of bamboo- and wood-based shear walls in light-frame buildings. *Energy & Buildings* 168: 167–179.

Westoby, M.J., Brasington, J., Glasser, N.F., Hambrey, M.J. & Reynolds, J.M. 2012. "Structure-from-Motion" photogrammetry: A low-cost, effective tool for geoscience applications. Geomorphology.

Wu, C. 2013. Towards linear-time incremental structure from motion. Pages 127–134 of: International Conference on 3d Vision.

Wu, Y. & Xiao, Y. 2018. Steel and glubam hybrid space truss. *Engineering Structures* 171:140–153.

Xiao, Y., Yang, R.Z. & Shan, B. 2013. Production, environmental impact and mechanical properties of glubam. *Construction and Building Materials* 44: 765–7773.

Yilmaz, H.M., Yakar, M., Gulec, S.A. & Dulgerler, O.N. 2007. Importance of digital close-range photogrammetry in documentation of cultural heritage. *Journal of Cultural Heritage.*

Modern Engineered Bamboo Structures – Xiao, Li & Liu (eds)
© 2020 Taylor & Francis Group, London, ISBN 978-1-138-35185-1

General parametric design of a steel-glubam hybrid space truss

K. Ma
Zhejiang University-University of Illinois at Urbana Champaign Institute, Jiaxing, Zhejiang, China

Y. Xiao
Zhejiang University-University of Illinois at Urbana Champaign Institute, Jiaxing, Zhejiang, China
Nanjing Tech University, Nanjing, Jiangsu, China
Department of Civil Engineering, University of Southern California, Los Angeles, CA, USA

ABSTRACT: This paper introduces a parametric design method for a hybrid truss system composed of glued laminated bamboo (glubam) and steel. Experiments on determining material's physical and mechanical parameters were carried out first, on basis of which design stages from modeling, analysis, optimization to manufacturing are all rendered possible through parametric ways by defining corresponding parameters within one single platform - Grasshopper. By maximizing automation during the process, efficiency and extensibility are taken into consideration for possibly further, larger, and more complex design.

1 INTRODUCTION

1.1 *Glubam*

When comparing with traditional-material-based structures, there are several advantages in timber structures, such as material renewability, low energy consumption, easy manufacturing and assembling, outstanding anti-seismic performance and resistance to low temperature. Despite all these advantages, timber structure is not widely used in China due to the lack of wood resources. The paradoxical relationship between wood supply and demand of wood has been forcing people into looking for available alternative resources.

Bamboo shares many features and advantages with timber, and is even stronger in perspective of renewability and specific physical and mechanical properties. To make bamboo a qualified building material, processing methods similar to glued laminated timber (glulam) was invented, and glued laminated bamboo (glubam), was born (Xiao et al., 2008, 2013).

1.2 *General parametric design*

Parametric modeling is famous for its ability to efficiently deal with complex structures, especially those with inner regularity so that can be described via algorithm, i.e. truss, frame and grid shell.

Despite the fact that building design process actually includes different stages, e.g. modeling, analysis, optimization, manufacturing and construction, traditional parametric design is often limited to one of these stages, especially the modeling part. Discontinuity between stages makes parametric design not as powerful as it should be, as failure of parameter transmission can greatly slow down the whole design process. Moreover, important information lost during stage change makes structural optimization limited to within specific stage, and therefore hard to be realized.

This paper focuses on the combination of different design stages into one single parametric process -the general parametric design. By maximizing parameter passing and automation during the process, it is now possible to alter parameters even at the very end of design stage,

Table 1. Material properties of Steel.

Type	Tensile Strength (MPa)	Yield Strength (MPa)	Elastic Modulus (GPa)	Density (kg/m3)
Steel	410	≤ 245	206	7850

Table 2. Mechanical properties of glubam.

Type	In-plane tensile strength (MPa)	In-plane compressive strength (MPa)	Bending strength (MPa)	Elastic-modulus (GPa)	Density (kg/m3)
Glubam	83	68	111	10.1	737

thus making it convenient to optimize structure by adjusting parameters defined earlier according to those generated later. Efficiency and extensibility are taken into consideration during the whole process, for possibly further, larger, and more complex design.

2 MATERIAL PROPERTIES USED IN THE TRUSS

The hybrid space truss is constituted of steel pipes and glubam rods. To maximize material utilization, the steel pipes are used for bottom chords which mainly sustain tensile loads, while glubam rods which share a larger radius of gyration, are used for top and oblique rods which may sustain compressive loads (Wu and Xiao, 2018).

Experiments were carried out to determine the material's property. For steel pipes of the bottom chords, a universal testing machine was employed in order to test specimens according to Chinese specifications (GB/T 228.1, 2010; GB/T 2975, 1998). To measure the large deformation after the yield, an extensometer was used. See Table 1 for results.

Given the fact that no practical testing standards are available for glubam, ASTM standards (ASTM D143, 1994) and Chinese specifications (GB/T 1928, 2009) for timber material are employed with modification. Tension test, compression test and bending test were conducted accordingly to determine relevant parameters. See Table 2 for results.

3 GENERAL PARAMETRIC DESIGN

3.1 *Parametric platform*

General parametric design includes but is not limited to the process of modeling, analyzing, optimization and manufacturing, all of which are based on the modeling stage. (except for itself) Besides, modeling stage mostly often requires and generates the largest number of parameters. Considering all above, authors use the Grasshopper as the parametric design platform.

Grasshopper was developed as a Rhino plugin mainly focusing itself on the modeling process, which is why it is commonly used by architects. Its complete software development kit and application programming interface results in its excellent extensibility, creating chances that all design stages can be integrated within. Figure 1 shows a basic Grasshopper component example.

To clarify, parameters in Grasshopper go beyond their native meaning as a number. Basically, anything can be a parameter, including but not limited to an int, a double, a bool, a string, a point, a line, a geometry or even a whole parametric model.

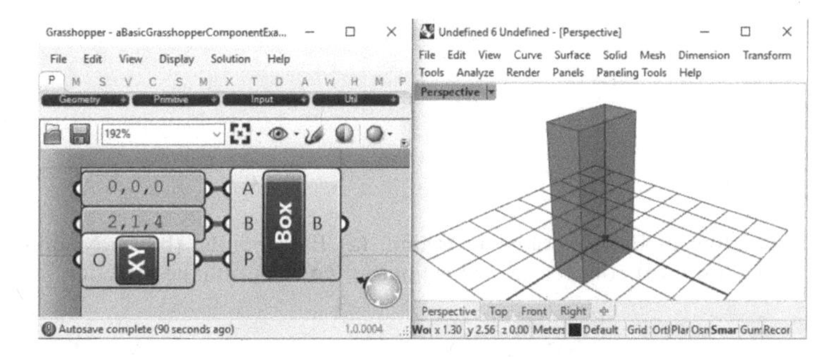

Figure 1. Left: Use a basic Grasshopper component 'Box 2pt' to define a Box with diagonal points (0,0,0) (2,1,4) and a base plane (XY plane). Right: Corresponding model generated in Rhino.

3.2 *Parametric modeling*

Most important geometry parameters of a truss structure are defined before and during this stage, including truss topology, grid dimension and member length. By default, authors set the dimension at 8 × 16, member length at 1.2 m on every direction. See Figure 2 for the whole process in Grasshopper.

To begin with, a row of nodes defining x-direction joints are generated with 'ArrLinear' component, with each joint assigned a unique index in the format of '(j)'. This row is then processed with 'ArrLinear' again to populate itself along y-direction so that a matrix of nodes is formed with each node assigned an index like '(i, j)'. This step generates all nodes of the bottom chord of the truss. Repeat this step to generate the top chord so that all nodes are generated, in the form of two matrixes, see Figure 2 (c) for example.

Write a C# component to connect nodes of all rows into line segments representing all rods in y-direction, then flip the matrix and do again for x-direction. Later, use similar methods to connect the top chord with the bottom ones in all four directions to form the oblique chords. Finally, sort all chords for output.

Additional modeling effects together with their parameters can be added to the modeling process without a rerun. For example, to spring the truss along y-direction. Simply write a C#

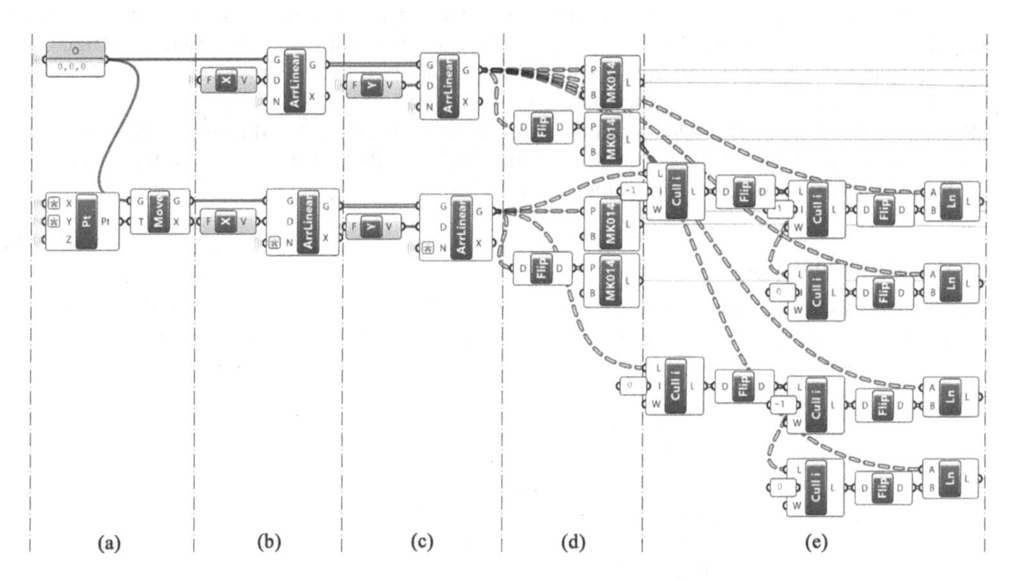

Figure 2. Modeling process example in Grasshopper: (a) Base points; (b) array along x-direction; (c) array along y-direction; (d) connect x&y-direction; (e) connect oblique-direction.

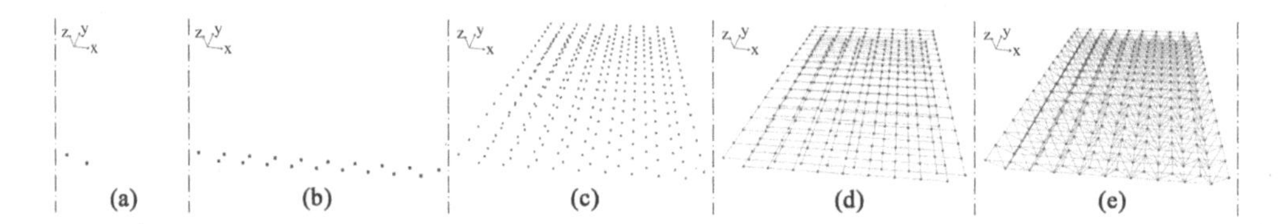

Figure 3. Corresponding model generated in Rhino: (a) Base points; (b) array along x-direction; (c) array along y-direction; (d) connect x&y-direction; (e) connect oblique-direction.

component to calculate the displacement vectors of each nodes to form a rational arch axis, and add a 'move' component to move all nodes along the vectors before the connection process for the spring to take effect.

3.3 *Parametric structural analysis*

Traditionally when doing structural analysis, parameters such as materials configurations and section assignments are defined within FEA software manually. Such cross-platform design experience can hardly promise the consistence of parameters, i.e. a small change in the modeling process may lead to a complete rebuild of the structural model, which significantly impairs design efficiency. Moreover, discontinuity between platforms restricts automated optimization due to a lack of feedback system.

To solve the across-platform-parameter-passing problem, two approaches can be adopted either establish a standard interface among platforms, or somehow manage to merge the platforms into a single one. On one hand, Grasshopper plugin GeometryGym was developed according to the Industry Foundation Classes (ISO, 2013), which enabling it to exchange modeling data among platforms, e.g. SAP2000 which supports IFC by default. On the other hand, plugin Karamba was developed as a FEA plugin working completely within the Grasshopper platform.

Simple supports are deployed symmetrically at the second node in y-direction of four corners of bottom layer (see Figure 5), and the loads are listed in Table 3 and combined according to Chinese load code (GB 50009, 2012).

Traditional finite element analysis software like SAP2000 excels in stability, and combination with parametric model makes it much stronger in perspective of modeling efficiency. Considering this, authors establish and analyze the same model through following methods:

- Modeled in Grasshopper and auto-exported to be analyzed in SAP2000.
- Modeled and analyzed in Grasshopper.

Results derived from the two methods agree with each other, validating the reliability of analysis outcome. Nevertheless, the second method has its advantage of better efficiency, which is why authors took it for parametric optimization.

3.4 *Parametric optimization*

In reference to Chinese specification on timber structure designing (GB 50005, 2003), maximum compressive capacity can be accordingly calculated as following

Table 3. Design loads.

Load effect	Value
Dead load	0.6 kN/m2
Wind load	−1.36 kN/m2
Snow load	0.65 kN/m2
Live load	0.5 kN/m2
Temperature Effect	+27°C; −16 °C

$$\varphi = \begin{cases} \frac{1}{1+\left(\frac{\lambda}{80}\right)^2} & \lambda \leq 75 \\ \frac{3000}{\lambda^2} & \lambda > 75 \end{cases} \tag{1}$$

$$F_{\max} \leq \varphi f A \tag{2}$$

where λ is the slenderness of compressive members, A is the cross-sectional area, φ is the stability coefficient, and f is glubam's mean In-plane compressive strength.

Considering the glubam rods all share square cross-section, φ and F_{max} depends solely on A, and thus can be deemed as functions of A. Therefore, to optimize the section size is the same to find the smallest A which satisfy

$$f(A) = \varphi(A)fA - F_{max}(A) \geq 0 \tag{3}$$

and to optimize the structure is the same to find the smallest quantity of materials

$$V = \sum nA_n l_n \tag{4}$$

where V is material total volume, An and ln are the cross-sectional area and length of the nth member. The optimized result can be acquired automatically, e.g. by using Bound Optimization BY Quadratic Approximation (BOBYQA) method for local optimization, or evolution algorithm for global one. For instance, authors managed to connect a C# Genetic Algorithm Library 'GeneticSharp' with grasshopper, as a customizable solver.

By applying members of different cross-section sizes along the truss and run multiple factor optimization, the structure can be further optimized. The optimization process is programmed to be able to handle multiple cross-section sizes. Nevertheless, considering the convenience of manufacturing and construction, authors set only two different cross-section-sizes at mid-span and near support respectively, divided along x&y-direction. The best dividing points are also deemed as parameters to be optimized, and to this specific structure are calculated at 35% (and symmetrically 65%) of total span in x-direction, and 24% (and symmetrically 76%) of total span in y-direction for best. With some C# component developed by authors, the convergence progress is recorded as Figure 4. See Figure 5 and Table 4 for optimization result detail.

3.5 *Parametric manufacturing and architectural rendering*

This part focuses mainly on automatic result output, both for mechanical processing and architectural purposes. Firstly, members are assigned with their optimized cross-sections.

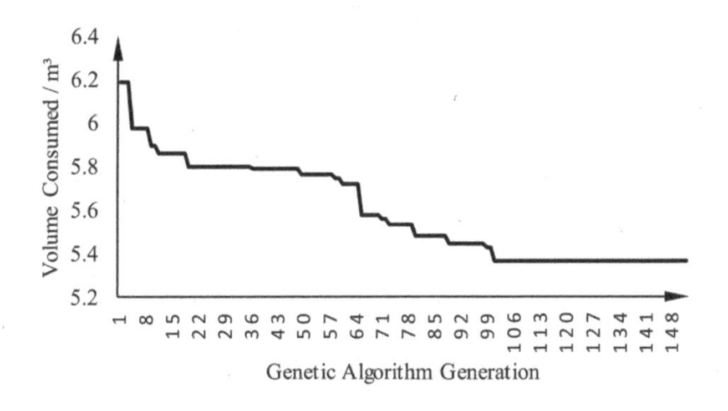

Figure 4. Genetic algorithm fitness convergence.

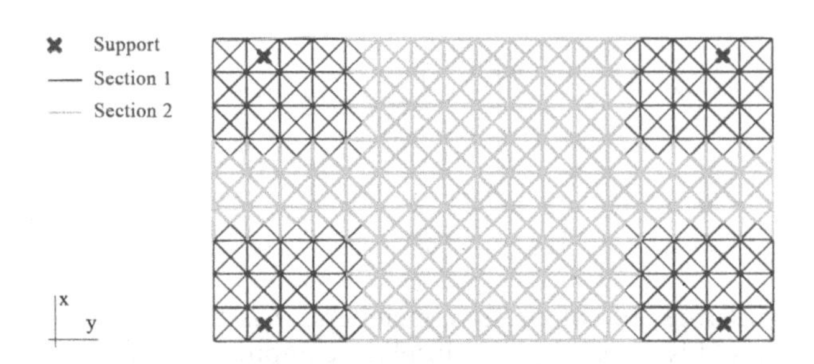

Figure 5. Optimization result of top and oblique chords.

Table 4. Optimized results comparison.

	Single section optimized	Two sections optimized
Total glubam volume	7.432 m^3	6.437 m^3
Optimized sections side length	81 mm	81 mm, 66 mm
Maximum Utilization	97.63%	98.63%, 98.05%

Rectangular cross-sections shall have their width sides parallel to horizontal plane. These components are then parametrically programmed by authors to be automatically pruned for joint engagement, including the following two steps:

- Create upright boxes using the components' axes as their diagonal lines. Preserve only the solid intersections of the components and their corresponding boxes, so as to ensure full engagement among all components, see Figure 6.
- Cut the components to create slots for steel joint plates and to automatically generate holes for bolts according to Chinese specifications (GB 50005, 2003), see Figure 6.

Secondly, use coordinates transformation to orient members into local system.

$$[T] = [\mathrm{O}'_{\mathrm{x}} \quad \mathrm{O}'_{\mathrm{y}} \quad \mathrm{O}'_{\mathrm{z}}] \tag{5}$$

$$\mathrm{A}' = [T]\mathrm{A} \tag{6}$$

where O' is unitized local system n-direction vector.

These 3-D models are then output to CAD and can be used for blueprint drawing, or even directly manufacturing by CAE, see Figure 6.

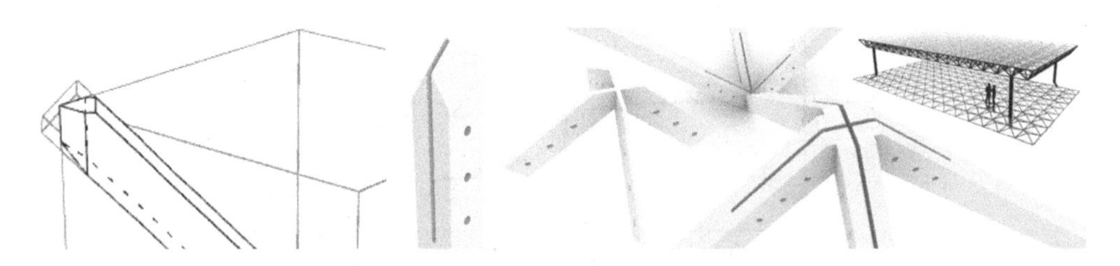

Figure 6. From left to right: Solid intersection; detailed model; gusset plate and joint; architectural rendering.

4 CONCLUSIONS

- The reasonability of using glubam and steel for hybrid space truss is validated via a whole set of standardized design process, including modeling, analysis, optimization and manufacturing.
- By integrating the whole design process within one parametric platform, the advantages of parametric design can be exploited as much as possible.
- Detailed instructions are given on how to establish the general parametric design program, based on which one can expect further, larger, and more complex future design.
- By using C# for secondary development of grasshopper, authors extended its ability to parametrically designing truss.

REFERENCES

ASTM D143 1994. Standard method of testing small clear specimens of timber.

GB 50005 2003. Code for design of timber structures. *Ministry of Construction of the People's Republic of China: Beijing, China.*

GB 50009 2012. Load code for the design of building structures. *Ministry of Housing and Urban-Rural Construction of the People's Republic of China, and General Administration of Quality Supervision, Inspection and Quarantine of the People's Republic of China: Beijing, China.*

GB/T 1928 2009. General requirements for physical and mechanical tests of wood.

GB/T 228.1 2010. Metallic materials-tensile testing-part 1: Method of test at room temperature. *the General Administration of Quality Supervision, Inspection and Quarantine of the People's Republic of China (AQSIQ) and the Standardization Administration (SAC) of the People's Republic of China: Beijing, China.*

GB/T 2975 1998. Steel and steel products-location and preparation of test pieces for mechanical testing. *State Quality and Technology Supervision Bureau: Beijing, China.*

ISO 2013. Industry foundation classes (IFC) for data sharing in the construction and facility management industries.

Wu, Y. & Xiao, Y. 2018. Steel and glubam hybrid space truss. *Engineering Structures* 171: 140–153.

Xiao, Y., Shan, B., Chen, G., Zhou, Q., & She, L.Y. 2008. Development of a new type glulam—glubam. In *Modern bamboo structures*, pp. 53–60. CRC Press.

Xiao, Y., Yang, R.Z., & Shan, B. 2013. Production, environmental impact and mechanical properties of glubam. *Construction and Building Materials* 44: 765–773.

Modern Engineered Bamboo Structures – Xiao, Li & Liu (eds)
© 2020 Taylor & Francis Group, London, ISBN 978-1-138-35185-1

Author Index